Health, safety and environment test

For managers and professionals

All the questions and answers for tests taken from April 2012

Published by CITB-ConstructionSkills, Bircham Newton, King's Lynn, Norfolk PE31 6RH

The Construction Industry Training Board otherwise known as CITB-ConstructionSkills and ConstructionSkills is a registered charity (Charity number: 264289)

First published 2006
Revised 2007, 2012

ISBN: 978-1-85751-342-4

CITB-ConstructionSkills has made every effort to ensure that the information contained within this publication is accurate. Its content should be used as guidance material and not as a replacement for current regulations or existing standards

Contents

F Specialist activities

Further information

Foreword

The construction industry will always be inherently dangerous for people working on site. It is the way we recognise and manage that risk which makes a difference to them.

The role of managers and professionals in helping to recognise and manage risk and so raise health and safety standards is significant and, until recently, underestimated. Whether they are architects, engineers, site managers, quantity surveyors or one of the many other disciplines involved in the design, procurement and construction process.

The CITB-ConstructionSkills' *Health, safety and environment test for managers and professionals* continues to give managers and professionals essential health, safety and environmental knowledge. This has an enormous influence, not just on the construction techniques used, but also on the values and ethos of the sites themselves.

Managing health, safety and the environment is not about policing. It concerns behaviour and the broad and deep consensus that unsafe sites or uncontrolled activities with the potential to damage the environment are unacceptable in any circumstances. This is a responsibility placed on all of us.

As professionals, we need to show leadership by example and not be escorted visitors on our own sites. Obtaining the correct competency card is a clear sign that a fully trained workforce is key.

We have a great industry. However, it must become a healthier and safer one for all those that work in it. It must also take its environmental responsibilities seriously for the sake of future generations.

Graham Watts OBE

Chief Executive
Construction Industry Council

About the test

The CITB-ConstructionSkills' Health, safety and environment test helps raise standards across the industry. It ensures that workers meet a minimum level of health, safety and environmental awareness before going on site. It should be used as a stepping stone, encouraging employers and their workforce to go on and develop their knowledge even further.

The test has been running for over 10 years and it underwent significant improvements in early 2012. These improvements included an enhanced test and a new delivery infrastructure.

All tests last for 45 minutes and have 50 multiple-choice questions including:

- 12 behavioural case study questions about how you should behave on a construction site to stay healthy and safe, followed by
- 38 knowledge questions to check your knowledge of health, safety and environment issues.

Different tests have been developed to meet the demands of different trades and professions. You will need to make sure that you take the right one.

The following tests are available:

- operative test
- specialist tests
- managers and professionals test.

What has changed in the new Health, safety and environment test?

New content has been introduced including:

- behavioural case studies
- knowledge questions on respiratory risks
- knowledge questions on environment
- a new specialist test for tunnelling
- a new structure for question selection.

What specialist tests are there?

There are currently 11 different specialist tests available in the CITB-ConstructionSkills' *Health, safety and environment test for operatives and specialists*. These include supervisors; demolition; plumbing or gas; highway works; specialist working at height; lifts and escalators; tunnelling; heating, ventilation, air conditioning and refrigeration (HVACR).

What are the behavioural case study questions?

The behavioural case study questions are designed to test how you respond to health and safety situations on a construction site. They are based on the principles established in the film *Setting out – what you should expect from a site and what they expect from you.*

Further information on the film and the transcript is provided at the back of this book.

Every test includes three case studies, each of which has four linked multiple–choice questions. These progress through a fictional situation faced by an individual working in the construction industry.

What are the knowledge questions?

The knowledge questions cover 16 core areas that are included in all the tests. These questions are very factual. The managers and professionals test also has a specialist knowledge question bank that covers specific areas (such as CDM regulations, demolition, plumbing or gas, and highway works).

You will not need a detailed knowledge of the exact content or working of any regulations. However, you will need to show that you know what is required of you, the things you must do (or not do), and what to do in certain circumstances (for example, discovering an accident).

Many of the questions refer to the duties of employers. In law, the self-employed can have the same legal responsibilities as employers. To keep the questions as brief as possible, the content only refers to the duties of employers but the questions apply to both.

The questions are based on British legislation. However, Northern Ireland legislation differs from that in the rest of the UK. For practical reasons, all candidates (including those in Northern Ireland) will be tested on questions using legislation relevant to the rest of the UK only.

What is a multiple-choice question?

The test is made up of multiple-choice questions. This means you will need to select the answer you think is correct from a set of possible answers.

Most questions will ask you to select a single answer. However some of the knowledge questions will ask you to select multiple answers. This will always be clearly stated in the question.

Who writes the questions?

The question bank is developed by CITB-ConstructionSkills with industry-recognised organisations which sit on or support the Health and Safety Test Question Sub-Committee. A full list of those parties that support the test is set out in the acknowledgements at the back of the book.

Will the questions change?

The Health, safety and environment test questions and answers change from time to time in accordance with legislation, regulations and best practice. CITB-ConstructionSkills makes every effort to keep the test and the revision material up to date.

- You will not be tested on questions that are no longer deemed to be appropriate by the Health and Safety Test Question Sub-Committee, perhaps due to legislation or regulatory changes.
- You will not be tested on knowledge questions that do not feature in the most up-to-date published book.

Preparing for a test

There are a number of ways you can prepare for your test.

	Operatives	Specialists	Supervisors	Manager
Watch *Setting out*	Free to view at *www.cskills.org/settingout*			
Read the question and answer books	*HS&E test for operatives and specialists – All the questions and answers* (GT 100)			*HS&E test for managers and professionals – All the questions and answers* (GT 200)
Use the revision DVD	*HS&E test for operatives and specialists – DVD* (GT 100 DVD)			*HS&E test for managers and professionals – DVD* (GT 200 DVD)
Read supporting knowledge material	*Safe start* (GE 707)	*Safe start* (GE 707) plus sector recommended supporting material	*Site supervision simplified* (GE 706)	*Construction site safety* (GE 700)
Complete an appropriate training course	Site Safety Plus – one-day Health and safety awareness course	Contact your industry body for recommendations	Site Safety Plus – two-day Site supervisors safety training scheme	Site Safety Plus – five-day Site managers safety training scheme

How can I increase my chances of success?

- Revise using the recommended learning aids, working through all the knowledge questions.
- Watch the *Setting out* film.
- Complete a recommended training course.
- Book your test when you are confident with your topics and questions.
- Complete a timed simulated test a few days before (available on the DVD).

What's on the DVD?

The DVD holds an interactive package that includes:

- the *Setting out* film and a sample behavioural case study
- all the knowledge questions and answers in both book and practice formats
- the real test tutorial
- a test simulator – all the functionality of the test with the real question bank
- voice-overs in English and Welsh for all questions
- the operative questions and *Setting out* are provided with voice-overs in the following languages:

  Bulgarian, Czech, German, Hungarian, Lithuanian, Polish, Portuguese, Punjabi, Romanian, Russian.

  Please note: British Sign Language assistance is not included on any of the DVDs.

Where can I buy the revision books and DVDs?

CITB-ConstructionSkills has developed a range of revision material, including question and answer books and DVDs that will help you to prepare for the test. For further information and to buy these products you can:

 go online at *www.cskills.org/hsanderevision*

 telephone 0344 994 4488
lines open Monday to Friday 8am to 8pm, and Saturday 8am to noon

visit a good bookshop, either in the high street or online.

Where can I find more information on your supporting publications?

CITB-ConstructionSkills has developed a range of publications that present a detailed and comprehensive guide to the full range of topics covered in the test. They can be used to build awareness and understanding of the issues surrounding health and safety on a construction site. They provide the context for the questions that are asked.

The range of products includes: *Safe start* (GE 707), *Site supervision simplified* (GE 706) and *Construction site safety* (GE 700). For further information and to buy these products you can:

 go online at *www.cskills.org/publications*

telephone 0344 994 4122
lines open Monday to Friday 8.30am to 5.30pm.

What is the Site Safety Plus Scheme?

The CITB-ConstructionSkills' Site Safety Plus Scheme is a comprehensive health and safety training programme designed to provide the building, civil engineering and allied industries with a range of courses for individuals seeking to develop their skill set in this area.

They are designed to give everyone from operative to senior manager the skill set they need to progress through the industry. From a one-day Health and safety awareness course to the five-day Site management safety training scheme (SMSTS), these courses will ensure that everyone benefits from the best possible training.

For further information you can:

 go online at *www.cskills.org/sitesafetyplus.*

Booking a test

The easiest way to book your test is either online or by telephone. You will be given the date and time of your test immediately and offered the opportunity to buy revision material, such as books and DVDs. You should be able to book a test at your preferred location within two weeks.

To book your test you can:

go online at *www.cskills.org/hsandetest*

telephone 0344 994 4488
lines open Monday to Friday 8am to 8pm, and Saturday 8am to noon.

post in an application form (application forms are available from the website and the telephone number listed above).

When booking your test you will be able to choose whether to receive confirmation by email or by letter. It is important that you check the details (including the type of test, the location, the date and time) and follow any instructions it gives regarding the test.

You can also choose to receive an SMS text message or email reminder 24 hours before your test.

For those instances where a test is required at short notice it may be possible to turn up at a centre and take a test on the day (subject to available spaces). It's strongly advised that you do not rely on this option.

What information do I need to book a test?

To book a test you should have the following information to hand:

- which test you need to take
- whether you require any additional assistance
- your chosen method of payment (debit or credit card details)
- your address details
- your CITB-ConstructionSkills' registration number (you will have one of these if you have taken the test before, or applied for certain card schemes including CSCS, CPCS, CISRS, etc.).

Where can I take a test?

To sit a Health, safety and environment test you will need to visit a CITB-ConstructionSkills' approved test centre. There are three different types of centre:

- a fixed test centre operated by our delivery provider.
  To find your nearest test centre visit *www.cskills.org/hsandetest.*
  These test centres will be able to offer tests from Monday to Saturday, but local opening times will vary. Normal opening hours are Monday to Friday 8am to 8pm and Saturdays 8am to noon
- an independent Internet Test Centre (for example as operated by a college, training provider or commercial organisation). Please contact these test centres directly for further information and to book a test

- a corporate testing unit which can be established at a suitable venue for a group of candidates. For further information on this service there is a dedicated booking line 0344 994 4492.

Is there any special assistance available when taking the test?

Voice-over assistance: All tests can be booked with English and Welsh voice-overs.

Foreign language assistance:

- The operatives test can be booked with voice-overs in the following languages: Bulgarian, Czech, German, Hungarian, Lithuanian, Polish, Portuguese, Punjabi, Romanian, Russian.

  An interpreter can be requested if assistance is required in other languages.
- The specialist tests can be booked with an interpreter but no pre-recorded voice-overs are available.
- The managers and professionals test does not allow foreign language assistance because a basic command of English or Welsh is required in order to sit the test.

Sign language assistance: The operatives test can be booked with British Sign Language on screen. If you need assistance in the other tests a signer can be provided.

Further assistance: If you need any other special assistance (such as a reader, signer, interpreter, or extra time) this can be provided but you will need to book through a dedicated booking line 0344 994 4491.

What services are there for Welsh speakers?

- All tests can be booked with Welsh voice-overs.
- All revision DVDs include Welsh voice-overs.
- There is a dedicated Welsh booking line 0344 994 4490.

How do I cancel or postpone my test?

To cancel or reschedule your test you should go online or call the booking number at least 72 hours (three working days) before your test, otherwise you will lose your fee.

What if I do not receive a confirmation email/letter?

If you do not receive a confirmation email or letter within the time specified please call the booking line to check your booking has been made.

We cannot take responsibility for postal delays. If you miss your test event, you will unfortunately forfeit your fee.

Taking a test

Before the test

On the day of the test you will need to:

- allow plenty of time to get to the test centre and arrive at least 15 minutes before the start of the test
- take your confirmation email or letter
- take proof of identity that bears your photo and your signature (such as driving licence card or passport – please visit *www.cskills.org/hsandetest* for full list of acceptable documentation).

On arrival at the test centre, staff will check your documents to ensure you are booked onto the correct test. If you do not have all the relevant documents you will not be able to sit your test and you will lose your fee.

During the test

The tests are all delivered on a computer screen. However, you do not need to be familiar with computers and the test does not involve any writing. All you need to do is click on the relevant answer boxes, using either a mouse or by touching the screen.

Before the test begins you can choose to work through a tutorial. It explains how the test works and lets you try out the buttons and functions that you will use while taking your test.

The test will contain 50 multiple-choice questions which you will need to complete in 45 minutes.

There will be information displayed on the screen which shows you how far you are through the test and how much time you have remaining.

After the test

At the end of the test there is an optional survey that gives you the chance to provide feedback on the test process.

You will be provided with a printed score report after you have left the test room. This will tell you whether you have passed or failed your test, and give feedback on areas where further learning and revision is recommended.

What do I do if I fail?

If you fail your test, your score report will provide feedback on areas where you got questions wrong.

It is strongly recommended that you revise these areas thoroughly before re-booking. You will have to wait at least 72 hours before you can take the test again.

What do I do if I pass?

A Health, safety and environment test pass is often a necessary requirement when applying to join a construction industry card scheme. Different schemes exist in different trades and professions. Membership of a relevant scheme helps you prove that you can do your job, and that you can do it safely. Access to construction sites may require a relevant scheme card.

Once you have passed your test, you should, if you have not done so already, consider applying to join the relevant card scheme. However please be aware that you may need to complete further training, assessment and/or testing to meet their specific entry requirements.

To find out more about many of the recognised schemes you can:

 go online at at *www.cskills.org/cardschemes.*

Further scheme contact details		telephone	go online
General	Construction Skills Certification Scheme (CSCS)	0844 576 8777	www.cscs.uk.com
	Northern Ireland: Construction Skills Register (CSR)	028 9087 7150	www.cefni.co.uk
Plant operatives	Construction Plant Competence Scheme (CPCS)	0844 815 7274	www.cskills.org/cpcs
Demolition operatives	Certificate of Competence of Demolition Operatives (CCDO)	0844 826 8385	www.ndtg.org
Scaffolders	Construction Industry Scaffolders Record Scheme (CISRS)	0844 815 7223	www.cisrs.org.uk
HVACR operatives	Engineering Services SKILLcard (ESS)	01768 860 406	www.skillcard.org.uk
Plumbers	Joint Industry Board for Plumbing and Mechanical Service (JIBPMES)	01480 476 925	www.jib-pmes.org
	Scottish and Northern Ireland Joint Industry Board (SNIJIB)	0131 225 2255	www.snijib.org
Electricians	Electrotechnical Certification Scheme (ECS)	In England call 0844 847 5098 In Scotland call 0131 445 5577	www.ecscard.org.uk

A
Legal and management

1.1

A whole site has been issued with a prohibition notice. During the period that the notice applies, what does this mean?

- A The site manager should be on site before work starts
- B The site manager must check with the Health and Safety Executive (HSE) before starting work
- C No-one must use any survey equipment, tools or machinery
- D All work must stop on site until the safety problem is rectified

1.2

Why is the Health and Safety at Work Act important to anyone at work? Give TWO answers.

- A It explains how health and safety is managed on site
- B It explains how to write risk assessments
- C It requires all employers to provide a safe place to work
- D It sets out how work should be carried out
- E It puts legal duties on workers with regard to their acts or omissions

1.3

It is important for those at work to see their employer's health and safety policy because it tells them:

- A how to do their job safely
- B the contents of the risk assessments
- C how health and safety is managed within their organisation
- D how to use tools and equipment safely

1.4

If a prohibition notice is issued by an inspector of the Health and Safety Executive (HSE) or local authority:

- A work can continue, provided that a risk assessment is carried out
- B the work that is subject to the notice must cease
- C the work can continue if extra safety precautions are taken
- D the work in hand can be completed, but no new works started

1.5

An employer has to prepare a written health and safety policy and record the significant findings of risk assessments when:

- A they employ three people or more
- B they employ five people or more
- C they employ 10 or more people
- D the work is going to last more than 30 days

Answers: 1.1 = D 1.2 = C, E 1.3 = C 1.4 = B 1.5 = B

1.6

If there is a fatal accident or reportable dangerous occurrence on site, when must the Health and Safety Executive (HSE) be informed?

- A Immediately
- B Within five days
- C Within seven days
- D Within 10 days

1.7

During site induction you do not understand something the presenter says. What should you do?

- A Attend another site induction
- B Ask the presenter to explain the point again
- C Guess what the presenter was trying to tell you
- D Wait until the end then ask someone else to explain

1.8

Now that work on site has moved forward, the safety rules given in your site induction seem out of date. What should you do?

- A Do nothing, you are not responsible for safety on site
- B Speak to the site manager about your concerns
- C Speak to your colleagues to see if they have any new rules
- D Decide yourself what to do to suit the changing conditions

1.9

If you discover children playing on site, what is the first priority?

- A Tell the site manager

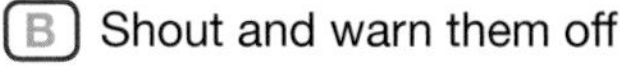

- B Shout and warn them off
- C Make sure the children are taken to a place of safety
- D Find out how they got into the site

1.10

The standards of health and safety on a project site have noticeably declined. As the responsible professional what is the FIRST thing you should do to find out about the contractor's attitude to health and safety?

- A Review their health and safety inspection reports
- B Go out on site and look
- C Start a dialogue with the site manager

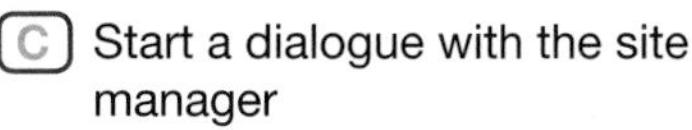

- D Call the contractor's safety department

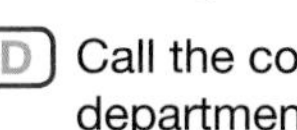

1.11

Why is it important that hazards are identified?

- A They have the potential to cause harm
- B They must all be eliminated before work can start
- C They must all be notified to the Health and Safety Executive (HSE)
- D They have to be written on the Health and Safety Law poster

Answers: 1.6 = A 1.7 = B 1.8 = B 1.9 = C 1.10 = C 1.11 = A

1.12

In the context of a risk assessment, what does the term 'risk' mean?

A Something with the potential to cause injury

B An unsafe act or condition

C The likelihood or chance that a hazard could actually cause harm or damage

D Any work activity that can be described as dangerous

1.13

Why may a young person be more at risk of having accidents?

A Legislation does not apply to anyone under 18 years of age

B They are usually left to work alone to gain experience

C They have less experience and may not recognise danger or understand fully what could go wrong

D There is no legal duty to provide them with personal protective equipment (PPE)

1.14

Two contractor's workers turn up on site with a file of generic risk assessments. You are keen for them to start but you know the risk assessments do not take into account the current site-specific hazards. What is your best course of action?

A Turn them away and tell them to come back with better paperwork?

B Get them to sign the generic risk assessments and tell them about the site

C Amend the risk assessments with them or their supervisor to take into account site specifics before they can start

D Cover the site-specific hazards during the site induction

1.15

A planned task that involves significant risk should only be undertaken by a:

A competent person

B worker

C supervisor

D health and safety professional

Answers: 1.12 = C 1.13 = C 1.14 = C 1.15 = A

1.16

The significant findings of a risk assessment reveal a risk to health or safety of site staff. What measure should always be considered first?

A) Make sure personal protective equipment (PPE) is available

B) Adapt the work to the individual

C) Give priority to those measures that protect the whole workforce

D) Avoid the risk altogether if possible

1.17

In considering what measures to take to protect people against risks to their health and safety, personal protective equipment (PPE) should always be considered:

A) first, because it is an effective way to protect people

B) as the only practical measure

C) as the best way to tackle the job

D) only when the risks cannot be eliminated by other means

1.18

What is the purpose of an on-site risk assessment?

A) To save time completing documentation

B) To review the actual hazards and risks and to ensure that any planned safety system is still applicable

C) To ensure that the work can be carried out in a reasonable timeframe

D) To protect the employer from prosecution

1.19

A risk assessment tells you:

A) how significant risks are being created

B) what legislation should be applied to control risks

C) the generic risks associated with an activity

D) what risks may exist and how they should be controlled

1.20

You will often hear the word hazard mentioned. What does it mean?

A) Anything that has the potential to cause harm or damage

B) The level of risk on site

C) A type of barrier or machine guard

D) All of these answers

Answers: 1.16 = D   1.17 = D   1.18 = B   1.19 = D   1.20 = A

1.21

As a summary of the general principles of prevention when developing safe methods of work ERIC stands for:

- (A) Employ, reduce, inform, control
- (B) Ensure, reduce, inform, control
- (C) Eliminate, reduce, inform, control
- (D) Educate, reduce, inform, control

1.22

Which of the following has the highest priority under the principles of prevention and protection?

- (A) Specifying the use of a mobile elevating work platform (MEWP) for steel erection
- (B) Pre-fabrication of steelwork that eliminates the need for a high-level bolted connection
- (C) The installation of a 2 m high barrier to prevent unauthorised access to an area where work at height is taking place
- (D) The introduction of a comprehensive briefing sheet to warn workers of the dangers of working at height

1.23

What is the purpose of using a 'permit-to-work' system?

- (A) To ensure that the job is being carried out properly
- (B) To ensure that the job is carried out by the easiest method
- (C) To enable tools and equipment to be properly checked before work starts
- (D) To establish a safe system of work

1.24

The number of people who may be carried in a passenger hoist on site must be:

- (A) displayed on a legible notice within the site welfare area
- (B) displayed on a legible notice within the cage of the hoist
- (C) given in the company safety policy
- (D) explained to the operator of the hoist

1.25

Welding is about to start on the site you are visiting. What should be provided to prevent passers-by from getting arc eye?

- (A) Warning signs
- (B) Screens
- (C) Personal protective equipment (PPE)
- (D) Nothing

Answers: 1.21 = C 1.22 = B 1.23 = D 1.24 = B 1.25 = B

1.26

From a safety point of view, which of the following should be considered first when deciding on the number and location of access and egress points to a site?

A Off road parking for cars and vans

B Access for the emergency services

C Access for heavy vehicles

D Site security

1.27

Which of the following can provide a good first impression of how well a site is currently being run?

A How tidy and organised it is

B What the contractor says in the monthly site meeting

C What the last health and safety inspection report says

D The number of signs displayed at the site entrance

1.28

On visiting a site you notice that it is adjacent to a primary school. What is likely to be the most effective way of keeping children off construction sites?

A Put up 'keep out' posters

B Erect security fencing or hoarding and keep all entrance gates closed

C Give safety talks to the local schools and youth clubs

D Send a flier to local households telling them to keep their children off site

1.29

Which of the following is a good reason for obtaining and communicating information on construction health risks?

A Many professionals and workers do not know enough about construction health risks

B Construction workers do not always understand the long-term effects on their health

C Ill health in construction workers is more common than traumatic injuries

D All of these answers

1.30

What is the best way for a responsible person to make sure that all who are doing a job have fully understood a method statement?

A Attach the method statement to the risk assessment and job sheet

B Explain the method statement to those doing the job and test their understanding

C Make sure that those doing the job have read the method statement

D Display the method statements on a notice board in the office

Answers: 1.26 = B 1.27 = A 1.28 = B 1.29 = D 1.30 = B

1.31

On what basis would you expect the topics for tool box talks to be selected?

A They are picked at random from the list of tool box talks

B So the topic relates to work that is being carried out at that time

C The client selects the topic for each talk

D In an order so that each topic is given at least once a year

Answers: 1.31 = B

2.1

Which of these does NOT have to be recorded in the accident book?

- A The injured person's national insurance number
- B The date and time of the accident
- C Details of the injury
- D The home address of the injured person

2.2

When must you record an accident in the accident book?

- A If you are injured in any way
- B Only if you have to be off work
- C Only if you have suffered a broken bone
- D Only if you have to go to hospital

2.3

If someone is injured at work, who should record it in the accident book?

- A The site manager and no-one else
- B The injured person or someone acting for them
- C The first aider and no-one else

- D Someone from the Health and Safety Executive (HSE)

2.4

If anyone has an accident at work it must be recorded. Accident records, which can be viewed by anyone, must:

- A contain the injured person's name and address
- B only be completed by a site manager or supervisor
- C comply with the requirements of the Data Protection Act
- D only be kept in an electronic format

2.5

Which sector of the construction industry generally has the highest fatal accident/incident rate?

- A Civil engineering sites
- B Demolition sites
- C Greenfield sites
- D Maintenance and refurbishment sites

2.6

What is the most important reason for keeping a working area on a construction site clean and tidy?

- A To prevent slips, trips and falls
- B So that the workers don't have to have a big clean-up at the end of the week

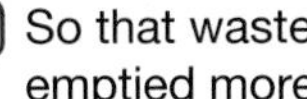

- C So that waste skips can be emptied more often
- D To recycle waste and help the environment

Answers: 2.1 = A 2.2 = A 2.3 = B 2.4 = C 2.5 = D 2.6 = A

2.7

Why is it important to attend site induction?

A You will get to know other new starters

B Risk assessments will be handed out

C Site-specific health and safety rules will be explained

D Permits to work will be handed out

2.8

Why is it important to report all accidents?

A It might stop them happening again

B Some types of accident have to be reported to the Health and Safety Executive (HSE)

C Details have to be entered in the accident book

D All of these answers

2.9

You can help prevent accidents by:

A reporting unsafe working conditions

B becoming a first aider

C knowing where the first-aid kit is kept

D knowing how to get help quickly

2.10

When are people, who are working on or visiting construction sites, most likely to have an accident?

A In the morning

B In the afternoon

C During the summer months

D When they first start on any new site

2.11

Historically, which type of accident kills most construction workers?

A Falling from height

B Contact with electricity

C Being run over by site transport

D Being hit by a falling object

2.12

Which of these helps everyone work safely on site?

A Site induction

B Tool box talks

C Risk assessments and method statements

D All of these answers

Answers: 2.7 = C 2.8 = D 2.9 = A 2.10 = D 2.11 = A 2.12 = D

2.13

Which TWO of the following can you learn from investigating an accident?

A That site operatives are always to blame

B That mechanical failures are most dangerous

C Ideas on how it could be prevented from happening again

D A combination of human error and mechanical failures always causes injury

E Details about why it happened

2.14

A scaffold has collapsed and you saw it happen. When you are asked about the accident, you should say:

A nothing, you are not a scaffold expert

B as little as possible because you don't want to get people into trouble

C exactly what you saw

D who you think is to blame and how they should be punished

2.15

A near miss is an incident where:

A you were just too late to see what happened

B someone could have been injured or something could have been damaged

C someone was injured and nearly had to go to hospital

D someone was injured and nearly had to take time off work

2.16

Why is it important to report near miss incidents on site?

A Because it is the law for all 'near miss' incidents

B To find someone to blame

C It is a requirement of the CDM Regulations

D To learn from them and stop them happening again

2.17

A member of the public has been injured by work activities. After emergency services have attended what should happen next?

A The accident should be reported to the Health and Safety Executive (HSE) or Incident Contact Centre

B The member of public should be told to keep quiet

C The member of public should be reminded to take greater care and attention near a site

D The accident should only be reported to the site health and safety manager

Answers: 2.13 = C, E 2.14 = C 2.15 = B 2.16 = D 2.17 = A

2.18

Who must be notified of a death, major injury, dangerous occurrence or over three-day injury on a site?

- (A) The company's insurance company
- (B) The local health centre
- (C) Either the Health and Safety Executive (HSE) or the Incident Contact Centre
- (D) No-one

2.19

An excavator on site has overturned but no-one was injured. What must happen next?

- (A) Clear up as quickly as possible and resume work
- (B) Investigate the dangerous occurrence
- (C) Make a report to the Health and Safety Executive (HSE) or Incident Contact Centre
- (D) Report the incident in the accident book

2.20

If you have a minor accident, who is responsible for making sure it is reported?

- (A) Anyone who saw the accident happen

- (B) A workmate
- (C) You
- (D) The Health and Safety Executive (HSE)

2.21

When leaving site you notice that a contractor is working in an unsafe manner. What should you do?

- (A) It is not your responsibility so leave site
- (B) Return and speak to the site manager
- (C) Contact the CDM co-ordinator
- (D) Contact the contractor's head office

2.22

If your doctor says that you contracted Weil's disease (leptospirosis) when on site, you will need to tell your employer. Why?

- (A) Your employer has to warn your colleagues not to go anywhere near you
- (B) Your employer has to report it to the Health and Safety Executive (HSE) as an occupational disease
- (C) Your work colleagues might catch it from you
- (D) The site on which you contracted it will have to be closed down

Answers: 2.18 = C 2.19 = B 2.20 = C 2.21 = B 2.22 = B

2.23

While on site you see a contractor working in a way that presents an imminent danger to yourself and others around you. What should you do immediately?

A Move to another area of the site and continue with your work

B Before you leave site ensure that you inform the site manager

C Speak directly to the site operatives and ask them to stop work and then tell the site manager

D Ensure that you inform the client and the CDM co-ordinator

Answers: 2.23 = C

3.1

You will find out about emergency assembly points from:

- A a risk assessment
- B a method statement
- C the site induction
- D the permit to work

3.2

How do you find out what to do if you are injured on site?

- A By asking someone on site
- B By looking for the first-aid sign
- C By attending a first-aid course
- D You should be told at site induction

3.3

In what way are site-based staff and visitors informed of the location of first-aid facilities on site?

- A By walking the site looking for the first-aid sign
- B By searching the site office
- C They should be told during site induction
- D By reading the Health and Safety Law poster

3.4

In which way should site visitors be informed of the actions to take in the event of an on-site emergency?

- A They should study the plans on the wall of the site office
- B They are informed during site induction
- C They should ask the site manager
- D They should take a look around the site for the emergency assembly point

3.5

How can you see for yourself that attention has been given to simple emergency procedures on site?

- A Scaffolding has inspection labels fitted
- B The distance between the structure and the assembly point is minimised
- C Fire points with extinguishers and a means of raising the alarm are in position
- D All electrical appliances have been electrically tested

3.6

Do those in charge of sites have to provide a first-aid box?

- A Yes, every site must have one
- B ONLY if more than 50 people work on site
- C ONLY if more than 25 people work on site
- D No, there is no legal duty to provide one

Answers: 3.1 = C 3.2 = D 3.3 = C 3.4 = B 3.5 = C 3.6 = A

3.7

How should you be informed about what to do in an emergency? Give TWO answers.

- A From the site induction
- B Look in the health and safety file
- C Ask the Health and Safety Executive (HSE)
- D Ask the local hospital
- E From the site notice boards

3.8

If there is an emergency while you are on site you should first:

- A leave the site and go back to your office
- B phone your office
- C follow the site emergency procedure
- D phone the police

3.9

An emergency route(s) must be provided and maintained at all times on construction sites to ensure safe passage to:

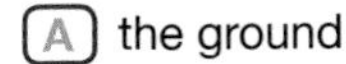

- A the ground
- B open air
- C a place of safety
- D the first-aid room

3.10

You have to carry out a survey on a remote unoccupied site. What should your employer provide you with?

- A A small first-aid kit
- B The first-aid box out of the office
- C Nothing
- D A book on first aid

3.11

When would you expect eyewash bottles to be provided?

- A Only on demolition sites where asbestos has to be removed
- B Only on sites where refurbishment is being carried out
- C On all sites where people could get something in their eyes
- D On all sites where showers are needed

3.12

If your employer's policy is that all staff who visit sites carry a travelling first-aid kit, it must NOT contain:

- A bandages
- B plasters
- C safety pins
- D over the counter medicines such as aspirin or painkillers

Answers: 3.7 = A, E 3.8 = C 3.9 = C 3.10 = A 3.11 = C 3.12 = D

3.13

If you cut your finger and it won't stop bleeding, you should:

- A wrap something around it and carry on working
- B tell the site manager
- C wash it clean then carry on working
- D find a first aider or get other medical help

3.14

Someone has fallen from height and has no feeling in their legs. You should:

- A roll them onto their back and keep their legs straight
- B roll them onto their side and bend their legs
- C ensure they stay still and don't move them until medical help arrives
- D raise their legs to see if any feeling comes back

3.15

If someone is in contact with a live cable the best thing you can do is:

- A phone the electricity company
- B dial 999 and ask for an ambulance
- C switch off the power and call for help
- D pull them away from the cable

3.16

What is the first thing you should do if you find an injured person?

- A Tell the site manager
- B Check that you are not in any danger before you check the injured person
- C Move the injured person to a safe place
- D Ask the injured person what happened

3.17

Someone working in a deep manhole has collapsed. What is the first thing you should do?

- A Get someone lowered into the manhole on a rope
- B Climb into the manhole and give mouth-to-mouth resuscitation
- C Run and tell the site manager
- D Shout and raise the alarm as a trained rescue team will be needed

3.18

Someone collapses with stomach pain and there is no first aider on site. What should you do first?

- A Get them to sit down
- B Get someone to call the emergency services
- C Get them to lie down in the recovery position
- D Give them some painkillers

Answers: 3.13 = D 3.14 = C 3.15 = C 3.16 = B 3.17 = D 3.18 = B

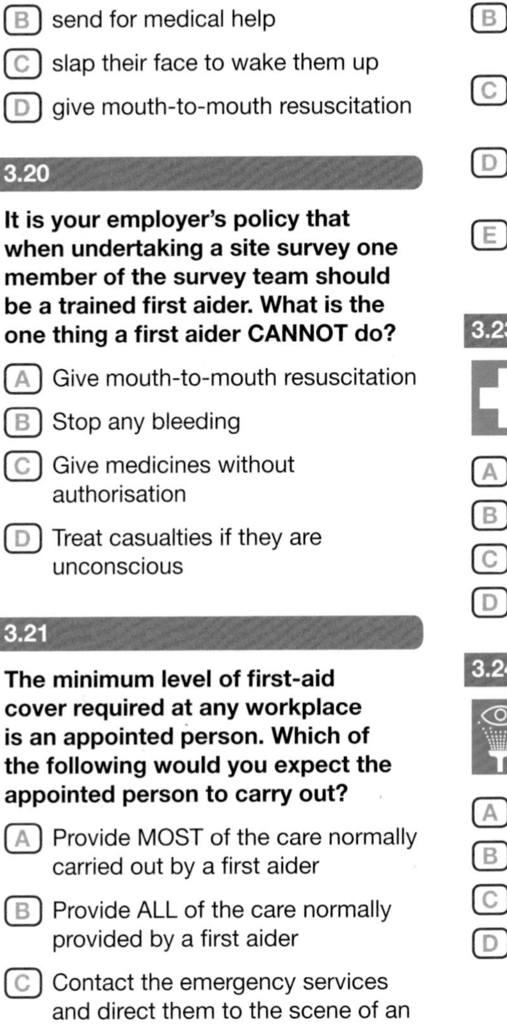

3.19

If someone falls and is knocked unconscious, you should first:

- A turn them over so they are lying on their back
- B send for medical help
- C slap their face to wake them up
- D give mouth-to-mouth resuscitation

3.20

It is your employer's policy that when undertaking a site survey one member of the survey team should be a trained first aider. What is the one thing a first aider CANNOT do?

- A Give mouth-to-mouth resuscitation
- B Stop any bleeding
- C Give medicines without authorisation
- D Treat casualties if they are unconscious

3.21

The minimum level of first-aid cover required at any workplace is an appointed person. Which of the following would you expect the appointed person to carry out?

- A Provide MOST of the care normally carried out by a first aider
- B Provide ALL of the care normally provided by a first aider
- C Contact the emergency services and direct them to the scene of an accident
- D Only apply plasters and dressings to minor wounds

3.22

Which TWO of the following factors must be considered when providing first-aid facilities on site?

- A The cost of first-aid equipment
- B The hazards, risks and nature of the work carried out
- C The number of people expected to be on site at any one time
- D The difficulty in finding time to purchase the necessary equipment
- E The space in the site office to store the necessary equipment

3.23

What does this sign mean?

- A First aid
- B Safe to cross
- C No waiting
- D Medicine box

3.24

What does this sign mean?

- A Safety glasses cleaning station
- B Emergency eyewash station
- C Warning, risk of splashing
- D Wear eye protection

Answers: 3.19 = B 3.20 = C 3.21 = C 3.22 = B,C 3.23 = A 3.24 = B

3.25

This sign tells you:

- (A) where the canteen is located
- (B) which way to walk
- (C) where to assemble in case of an emergency or evacuation
- (D) where the site induction room is located

3.26

If you think someone has a broken leg you should:

- (A) lie them on their side in the recovery position
- (B) use your belt to strap their legs together
- (C) send for the first aider or get other help
- (D) lie them on their back

3.27

If someone gets some grit in their eye, the best thing you can do is:

- (A) hold the eye open and wipe it with clean tissue paper
- (B) ask them to rub the eye until it starts to water
- (C) tell them to blink a couple of times
- (D) hold the eye open and flush it with sterilised water or eyewash

3.28

Someone gets a large splinter in their hand. It is deep under the skin and it hurts. What should you do?

- (A) Use something sharp to dig it out
- (B) Make sure they get first aid
- (C) Tell them to ignore it and let the splinter come out on its own
- (D) Try to squeeze out the splinter with your thumbs

3.29

Someone has got a nail in their foot. You are not a first aider. You must not pull out the nail because:

- (A) you will let air and bacteria get into the wound
- (B) the nail is helping to reduce the bleeding
- (C) it will prove that the casualty was not wearing safety boots
- (D) the nail would become a bio-hazard

3.30

If someone burns their hand the best thing you can do is:

- (A) put the hand into cold water or under a cold running tap
- (B) tell them to carry on working to exercise the hand
- (C) rub barrier cream or Vaseline into the burn
- (D) wrap your handkerchief around the burn

Answers: 3.25 = C 3.26 = C 3.27 = D 3.28 = B 3.29 = B 3.30 = A

B
Health and welfare

4.1

You can catch an infection called tetanus from contaminated land or water. How does it get into your body?

A Through your nose when you breathe

B Through an open cut in your skin

C Through your mouth when you eat or drink

D It doesn't, it only infects animals and not people

4.2

Your doctor has given you some medication. Which of these questions is the most important?

A Will it make me drowsy or unsafe to work?

B Will the medication make me work more slowly?

C Will my manager need to be told?

D Will it cause me to oversleep and be late for work?

4.3

Someone goes to the pub at lunchtime and has a couple of pints of beer. What should they do next?

A Drink plenty of strong coffee then go back to work

B Stay away from the site for the rest of the day

C Stay away for an hour and then go back to work

D Eat something, wait 30 minutes and then go back to work

4.4

Look at these statements about illegal drugs in the workplace. Which one is true in relation to site work?

A People under the influence of illegal drugs at work are a danger to everyone

B People who take illegal drugs work better and faster

C People who take illegal drugs take fewer days off work

D Taking illegal drugs is a personal choice so other people should not worry about it

4.5

White spirit or other solvents should not be used to clean hands because they:

A strip the protective oils from the skin

B remove the top layer of skin

C block the pores of the skin

D carry harmful bacteria that attack the skin

Answers: 4.1 = B 4.2 = A 4.3 = B 4.4 = A 4.5 = A

4.6

If you get a harmful substance on your hands, it can pass from your hands to your mouth when you eat. Give TWO ways to stop this.

A Wear protective gloves when you are working

B Wash your hands before eating

C Put barrier cream on your hands before eating

D Wear protective gloves then turn them inside-out before eating

E Wash your work gloves then put them on again before eating

4.7

When visiting a site you find that there is nowhere for you to wash your hands. What should you do?

A Wait until you get home then wash them

B Go to a local café or pub and use the washbasin in their toilet

C Speak to the site manager about the problem

D Bring your own hand washing equipment in future

4.8

What is the minimum that should be provided on site for washing hands?

A Nothing, there is no need to provide washing facilities

B Running hot water and electric hand-dryers

C A cold water standpipe and paper towels

D Hot and cold water (or warm water), soap and a way to dry hands

4.9

Direct sunlight on bare skin can cause:

A dermatitis

B rickets

C acne

D skin cancer

4.10

Exposure to engine oil and other mineral oils can cause:

A skin problems

B heart disease

C breathing problems

D vibration white finger

4.11

You can get occupational dermatitis from:

A hand-arm vibration

B another person with dermatitis

C some types of strong chemical

D sunlight

Answers: 4.6 = A, B 4.7 = C 4.8 = D 4.9 = D 4.10 = A 4.11 = C

4.12

You should not rely just on barrier cream to protect your skin from harmful substances because:

- A there may be none available on site
- B many harmful substances go straight through it
- C it is difficult to wash off
- D it can irritate your skin

4.13

When site workers need to handle harmful substances, they should wear the correct protective gloves to help stop:

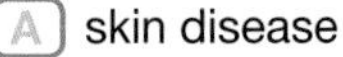

- A skin disease
- B vibration white finger
- C Raynaud's syndrome
- D arthritis

4.14

You are more likely to catch Weil's disease (leptospirosis) if you:

- A work near wet ground, waterways or sewers
- B work near air-conditioning units
- C work on building refurbishment
- D drink water from a standpipe

4.15

The early signs of Weil's disease (leptospirosis) can be easily confused with:

- A dermatitis
- B diabetes
- C hay fever
- D influenza (flu)

4.16

Which of the following species of animal is the most likely carrier of Weil's disease (leptospirosis) on construction sites?

- A Rabbits
- B Rats
- C Squirrels
- D Mice

4.17

You find that the toilets on site are dirty. What should you do?

- A Ignore the problem, it is normal
- B Make sure that you tell the site manager who can sort it out
- C Ask someone to find some cleaning materials and attend to it
- D See if you can use the toilets in a nearby café or pub

Answers: 4.12 = B 4.13 = A 4.14 = A 4.15 = D 4.16 = B 4.17 = B

4.18

Which of the following is a legal requirement under the CDM Regulations for the site welfare facilities?

A Wholesome drinking water, means to boil water and heat food

B Suitable facilities for rest, including tables and chairs with backs and, where necessary, changing rooms and lockers

C Suitable and sufficient toilets and washing facilities, including running cold and hot or warm running water

D All of these answers

4.19

The site toilets do not flush. What should you do?

A Try not to use the toilets while you are at work

B Tell the site manager about the problem

C Try to fix the fault yourself

D Ask a plumber to fix the fault

4.20

What sort of rest area should be provided for operatives on site?

A A covered area

B A covered area and some chairs

C A covered area, tables and chairs with backs, and something to boil water and heat food

D Nothing, contractors don't have to provide rest areas

Answers: 4.18 = D 4.19 = B 4.20 = C

5.1

Someone near you is using a disc-cutter to cut concrete blocks. What THREE immediate hazards are likely to affect you?

A Flying fragments

B Dermatitis

C Harmful dust in the air

D High noise levels

E Skin cancer

5.2

If using on-tool extraction to control dust from a power tool it is important to check that:

A the extraction unit is the correct type

B the extraction filters are clear and the unit is extracting dust

C you are using the power tool correctly

D All of these answers

5.3

When drilling, cutting, sanding or grinding you can breathe in high levels of harmful dust. These levels are likely to be highest when working:

A outside on a still day

B outside on a windy day

C in a small room

D in a large indoor space

5.4

You have finished your work and need to sweep up the dust. What should you do?

A Dampen down the area

B Make sure there is plenty of ventilation

C Put your protective mask back on

D All of these answers

5.5

You are using water as part of dust control and run out. Should you:

A carry on as you have nearly finished

B stop and refill with water

C ask everyone to clear the area and then carry on

D carry on but get someone to sweep up afterwards

5.6

When using power tools it is important to:

A stop dust getting into the air

B stand down wind of any dust

C do the work quickly to limit dust exposure

D only undertake the work during damp or wet weather

Answers: 5.1 = A, C, D   5.2 = D   5.3 = C   5.4 = D   5.5 = B   5.6 = A

5.7

You have been asked to do some work that will create dust. What should you do?

A You should not do the work. Dust is highly dangerous

B Use equipment which will eliminate or reduce the amount of dust whilst wearing correct personal protective equipment (PPE)

C Start work – no controls are needed as it's only dust

D Work for short periods at a time

5.8

When using water to keep dust down when cutting you must ensure:

A there is as much water as possible

B the water flow is correctly adjusted

C somebody stands next to you and pours water from a bottle

D all the water is poured on the surface to soak it before you start cutting

5.9

You need to use a power tool to cut or grind materials. Give TWO ways to control the dust.

A Work slowly and carefully

B Fit a dust extractor or collector to the machine

C Wet cutting

D Keep the area clean and tidy

E Wear a dust mask or respirator

5.10

If you use a power tool to cut or grind materials, why must the dust be collected and not get into the air?

A To save time and avoid having to clear up the mess

B Most dust can be harmful if breathed in

C The tool will go faster if the dust is collected

D You do not need a machine guard if the dust is collected

5.11

Pigeons' droppings and nests that can be hazardous to your health are found in an area where you are required to work. You should:

A carry on with your work carefully

B stop work and seek advice

C try to catch the pigeons

D let them fly away before carrying on with your work

5.12

Breathing in a dusty atmosphere for long periods can cause:

A occupational asthma

B occupational dermatitis

C skin cancer

D Weil's disease (leptospirosis)

Answers: 5.7 = B 5.8 = B 5.9 = B, C 5.10 = B 5.11 = B 5.12 = A

5.13

Occupational asthma can stop you working again with certain substances. It is caused by:

A exposure to loud noise

B exposure to rat urine

C skin contact with any hazardous substance

D breathing in hazardous dust, fumes or vapours

5.14

More construction workers die or suffer long term health issues from:

A falling from height

B being struck by a vehicle

C slipping and tripping

D breathing in hazardous substances

5.15

You have been given a dust mask to protect you against hazardous fumes. What should you do?

A Do not start work until you have the correct respiratory protective equipment (RPE)

B Do the job but work quickly

C Start work but take a break now and again

D Wear a second dust mask on top of the first one

5.16

Your workers have been face-fit tested for their respiratory masks. What is one of the checks you can make to ensure their mask fit is still effective?

A That they have no more than one day's beard/stubble growth

B That they are clean shaven

C That they have no more than one week's beard/stubble growth

D Stubble/beard has no effect on a mask's performance

5.17

Disposable masks have filtering face piece ratings of FFP1, FFP2 and FFP3. Which offers the greater protection for workers?

A FFP1

B FFP2

C FFP3

D They all offer the same protection – the numbers refer to the different sizes of mask

5.18

Which of the following do you need to do to ensure that your mask works?

A Check it's the correct type needed

B Pass a face-fit test wearing the mask

C Check you are wearing it correctly

D All of these answers

Answers: 5.13 = D 5.14 = D 5.15 = A 5.16 = A 5.17 = C 5.18 = D

5.19

Generally speaking how long can you use the same disposable mask for?

A Five working days

B Until it looks too dirty to wear

C One day or one shift

D 28 days

5.20

Which of these activities does NOT create silica dust, which is harmful if breathed in?

A Sawing timber and plywood

B Cutting kerbs, stone, paving slabs, bricks and blocks

C Breaking up concrete floors and screeds

D Chasing out walls and mortar joints or sweeping up rubble

5.21

The high levels of solvents in some paints and resins can cause:

A headaches, dizziness and sickness

B lung problems

C effects on other parts of your body

D All of these answers

5.22

When drilling, cutting, sanding or grinding what is the best way to protect your long term health from harmful dust?

A Use dust extraction or wet cut and wear light eye protection

B Wear a FFP3-rated dust mask and impact goggles

C Wear any disposable dust mask, hearing protection and impact goggles

D Use dust extraction or wet cut, wear a FFP3-rated dust mask, hearing protection and impact goggles

Answers: 5.19 = C 5.20 = A 5.21 = D 5.22 = D

6.1

Noise over a long time can damage your hearing. Can this damage be reversed?

- A Yes, with time
- B Yes, if you have an operation
- C No, the damage is permanent
- D Yes, if you change jobs

6.2

How can noise affect your health? Give TWO answers.

- A Headaches
- B Ear infections
- C Hearing loss
- D Waxy ears
- E Vibration white finger

6.3

After standing alongside noisy equipment, you have a 'ringing' sound in your ears. What does this mean?

- A Your hearing has been temporarily damaged
- B You have also been subjected to vibration
- C You are about to go down with the flu
- D The noise level was high but acceptable

6.4

Noise can damage your hearing. What is an early sign of this?

- A There are no early signs
- B Temporary deafness or ringing noise in your ears
- C A skin rash around the ears
- D Ear infections

6.5

When referring to noise, what does the term 'upper exposure action value' mean?

- A The level at which hearing protection zones must be established and hearing protection must be worn
- B The second time a noise reading is taken
- C The time at which a second pair of ear defenders are provided
- D When two noise meters are required

6.6

If you wear hearing protection, it will:

- A stop you hearing all noise
- B reduce noise to an acceptable level
- C repair your hearing if it is damaged
- D make you hear better

Answers: 6.1 = C 6.2 = A, C 6.3 = A 6.4 = B 6.5 = A 6.6 = B

6.7

If you need to wear hearing protection, you must remember that:

A you have to carry out your own noise assessment

B you have to pay for all hearing protection

C ear plugs don't work

D you may be less aware of what is going on around you

6.8

TWO recommended ways to protect your hearing are by using:

A rolled tissue paper

B cotton wool pads over your ears

C ear plugs

D soft cloth pads over your ears

E ear defenders

6.9

You need to wear ear defenders, but an ear pad is missing from one of the shells. What should you do?

A Leave them off and go on to site without any hearing protection

B Put them on and go on to site with them as they are

C Do not visit noisy areas on site until they are replaced

D Use an ear plug in one ear and then put them on

6.10

Someone near you is using noisy equipment and you have no hearing protection. What should you do?

A Ask them to stop what they are doing

B Carry on with your work because it is always noisy on site

C Leave the area until you have the correct personal protective equipment (PPE)

D Speak to the operative's supervisor

6.11

You have to inspect a site near a particular construction operation that is generating a high level of noise. It is not possible to shut the operation down. Which of the following actions would you expect to be the site manager's immediate response?

A Arrange for a noise assessment to be carried out

B Make hearing protection available to those people who ask for it

C Issue all people affected with hearing protection as a precaution

D Erect 'hearing protection zone' signs

Answers: 6.7 = D 6.8 = C, E 6.9 = C 6.10 = C 6.11 = C

6.12

If you have to enter in a 'hearing protection zone', you must:

- A not make any noise
- B wear the correct hearing protection at all times
- C take hearing protection with you in case you need to use it
- D wear hearing protection if the noise gets too loud for you

6.13

As a rule of thumb noise levels may be a problem if you have to shout to be clearly heard by someone who is standing:

- A 2 m away
- B 4 m away
- C 5 m away
- D 6 m away

6.14

What is the significance of the weekly or daily personal noise exposure limit value of 87 dB set out in the Control of Noise at Work Regulations?

- A All site personnel and visitors need to be warned if this noise level is being exceeded
- B Hearing protection needs to be provided upon request if this level is likely to be exceeded
- C The principal contractor must make sure everyone wears their hearing protection if this noise level is exceeded
- D Employers must ensure that their personnel are not exposed to noise above this level

6.15

If you need to wear disposable ear plugs how should you insert them so they protect your hearing from damage?

- A Only put them in when it starts getting very noisy
- B Only ever insert them half way into your ear
- C Roll them up and insert them as far as you can, while pulling the top of your ear up to open up the ear canal
- D Fold them in half and wedge them into your ear

6.16

Why is vibration a serious health issue?

- A There are no early warning signs
- B The long-term effects of vibration are not known
- C There is no way that exposure to vibration can be prevented
- D Vibration can cause a disabling injury that cannot be cured

6.17

What is vibration white finger?

- A A mild skin rash that will go away
- B A serious skin condition that will not clear up
- C Industrial dermatitis
- D A sign of damage to someone's hands and arms that might not go away

Answers: 6.12 = B 6.13 = A 6.14 = D 6.15 = C 6.16 = D 6.17 = D

6.18

Hand-arm vibration can cause:

- A skin cancer
- B skin irritation, like dermatitis
- C blisters to hands and arms
- D damaged blood vessels and nerves in fingers and hands

6.19

Who should the employer inform if someone reports to the site manager that they have work-related hand-arm vibration syndrome?

- A The Health and Safety Executive (HSE)
- B The local health authority
- C A coroner
- D The nearest hospital

6.20

Which of these is most likely to cause vibration white finger?

- A Handsaw
- B Hammer drill
- C Hammer and chisel
- D Battery-powered screwdriver

6.21

Operatives using machinery that can cause vibration are likely to suffer less from hand-arm vibration if they are:

- A very cold but dry
- B cold and wet
- C warm and dry
- D very wet but warm

6.22

What is the least reliable source of information when assessing the level of vibration from a powered percussive hand tool?

- A In-use vibration measurement of the tool
- B Vibration figures taken from the tool manufacturer's handbook
- C The judgement of the site manager based upon observation
- D Vibration data from the Health and Safety Executive's (HSE) master list

Answers: 6.18 = D 6.19 = A 6.20 = B 6.21 = C 6.22 = C

7.1

Which one of these is NOT a primary purpose of an asbestos survey?

A To provide accurate information on the location, amount and condition of asbestos materials

B To identify all asbestos materials that need to be removed before demolition or refurbishment work

C To help the management of any asbestos in a building

D To estimate how much it would cost to remove any asbestos

7.2

If you breathe in asbestos dust, it can cause:

A aching muscles and painful joints

B throat infections

C lung diseases

D dizziness and headaches

7.3

When visiting site the contractor thinks that they have found some asbestos, what is the first thing that should be done?

A Stop work and get everyone out of the affected area

B A sample should be taken to the site manager

C The bits should be put in a bin and work should carry on

D Find the first aider

7.4

Which of these statements applies to asbestos? It is:

A harmful to health

B fibrous mineral

C likely to be found in buildings built or refurbished before 2000

D all of these answers

7.5

You are visiting a site where an active asbestos removal enclosure has been set up. Which of the following would indicate that it is operating efficiently?

A Appropriate signage

B Everyone to be wearing red suits

C The sides of the enclosure bowing in

D Everyone is wearing respiratory protective equipment (RPE)

7.6

Which of these does NOT cause skin problems?

A Asbestos

B Bitumens

C Epoxy resins

D Solvents

Answers: 7.1 = D 7.2 = C 7.3 = A 7.4 = D 7.5 = C 7.6 = A

7.7

If asbestos is present what should happen before demolition or refurbishment takes place?

- A Advise workers that asbestos is present and continue with demolition
- B All asbestos should be removed as far as reasonably practicable
- C Advise the Health and Safety Executive (HSE) that asbestos is present and continue with demolition
- D Inspect the condition of the asbestos materials

7.8

What kind of survey is required to identify asbestos prior to any work being carried out on a pre-2000 building?

- A Type 3 survey
- B Management survey
- C Refurbishment and demolition survey
- D Type 2 survey

7.9

Where might you come across asbestos?

- A In a house built between 1950 and 1990
- B In any building built or refurbished before the year 2000
- C In industrial buildings built between 1920 and 1990
- D Asbestos has now been removed from all buildings

7.10

Cement bags have an additive to help prevent allergic dermatitis. When using a new bag what should be checked?

- A The bag is undamaged
- B The 'use by' date has not expired
- C It has been stored in a dry place
- D The contents are not hard and gone off

7.11

Why should you not kneel in wet cement, screed or concrete?

- A It will make your trousers wet
- B It is not an effective way to work
- C It can cause serious chemical burns to your legs
- D It will affect the finish

7.12

Wet cement, mortar and concrete are hazardous to your health as they cause:

- A dizziness and headaches
- B chemical burns and dermatitis
- C muscle aches
- D arc eye

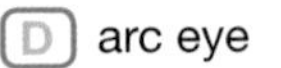

Answers: 7.7 = B 7.8 = C 7.9 = B 7.10 = B 7.11 = C 7.12 = B

7.13

You have to use a new material for the first time and need to carry out a COSHH assessment. What are the TWO main things you will need?

- A Your company's safety policy
- B The material safety data sheet
- C The age of the people doing the work
- D The material delivery note
- E Details of where, who and how you will be using the product

7.14

The safest way to use a hazardous substance is to:

- A get on with the job as quickly as possible
- B read your employer's health and safety policy
- C understand the COSHH assessment and follow the instructions
- D ask someone who has already used it

7.15

When a contractor is assessing the risk of using a substance they believe to be hazardous they should first:

- A review the material safety data sheet
- B ensure that safe storage is available on site
- C ensure workers are provided with respiratory equipment
- D ensure workers are trained to use respiratory equipment

7.16

The COSHH Regulations deal with:

- A the safe use of tools and equipment
- B the safe use of lifting equipment
- C the control and safe use of substances hazardous to health
- D safe working at height

7.17

Which of these will give you health and safety information about a hazardous substance?

- A The site diary
- B The delivery note
- C The COSHH assessment
- D The accident book

7.18

An assessment has been carried out under the COSHH Regulations. To which of the following should the risks and control measures be explained?

- A All who are working on site
- B Those on site using, or likely to be affected by, the substance
- C The person in charge of ordering materials
- D The accounts department

Answers: 7.13 = B, E   7.14 = C   7.15 = A   7.16 = C   7.17 = C   7.18 = B

7.19

Employers must prevent exposure of their workers to substances hazardous to health, where this is reasonably practicable. If it is not reasonably practicable, which of the following should be considered first?

A What instruction, training and supervision to provide

B What health surveillance arrangements will be needed

C How to minimise risk and control exposure

D How to monitor the exposure of workers in the workplace

7.20

How should cylinders containing liquefied petroleum gas (LPG) be stored on site?

A In a locked cellar with clear warning signs

B In a locked external compound at least 3 m from any oxygen cylinders

C As close to the point of use as possible

D Covered by a tarpaulin to shield the compressed cylinder from sunlight

7.21

Where should liquefied petroleum gas (LPG) cylinders be positioned when supplying an appliance in a site cabin?

A Inside the site cabin in a locked cupboard

B Under the cabin

C Inside the cabin next to the appliance

D Outside the cabin

7.22

You are visiting a site where flooring is being stuck down by a lone worker, using a liquid adhesive in a small inner room that has no visible means of ventilation. For what reason might you quickly bring this to the attention of the site manager?

A It is illegal for anyone to work on their own

B The work should be carried out under a hot work permit

C Kneeling and working is bad for their back

D The vapours from the adhesive may be a health hazard without sufficient fresh air

7.23

You find an unmarked container that you suspect may contain chemicals. What action should you take?

A Smell the chemical to see what it is

B Put it in a bin to get rid of it

C Move it to somewhere safe

D Ensure that it remains undisturbed and report it

Answers: 7.19 = C 7.20 = B 7.21 = D 7.22 = D 7.23 = D

7.24

You are visiting a project that involves removing paint from old iron work. Which of the following would enable the contractor to assess the foreseeable health risk of the work during the tender period?

(A) Lab-test results of a sample of paint giving lead content

(B) The prevailing wind conditions

(C) Fit testing of respiratory protective equipment (RPE)

(D) Tests to determine the average paint thickness

7.25

If you see either of these labels on a substance what should you do?

(A) Do not use it as the substance is poisonous

(B) Find out what protection you need as the substance is corrosive and can damage your skin upon contact

(C) Wash your hands after you have used the substance

(D) Find out what hand cleaner you will need as the substance will not wash off easily

7.26

If you see either of these labels on a substance what should you do?

(A) Find out what protection you need as the substance is harmful and could damage your health

(B) Use sparingly as the substance is expensive

(C) Wear gloves as the substance can burn your skin

(D) Do not use it as the substance is poisonous

7.27

If you see either of these labels on a substance what should you do?

(A) Make sure it is stored out of the reach of children

(B) Use the substance very carefully and make sure you don't spill or splash it on you

(C) Do not use it as the substance is poisonous

(D) Find out what protection you need as the substance is toxic and in low quantities could seriously damage your health or kill you

Answers: 7.24 = A 7.25 = B 7.26 = A 7.27 = D

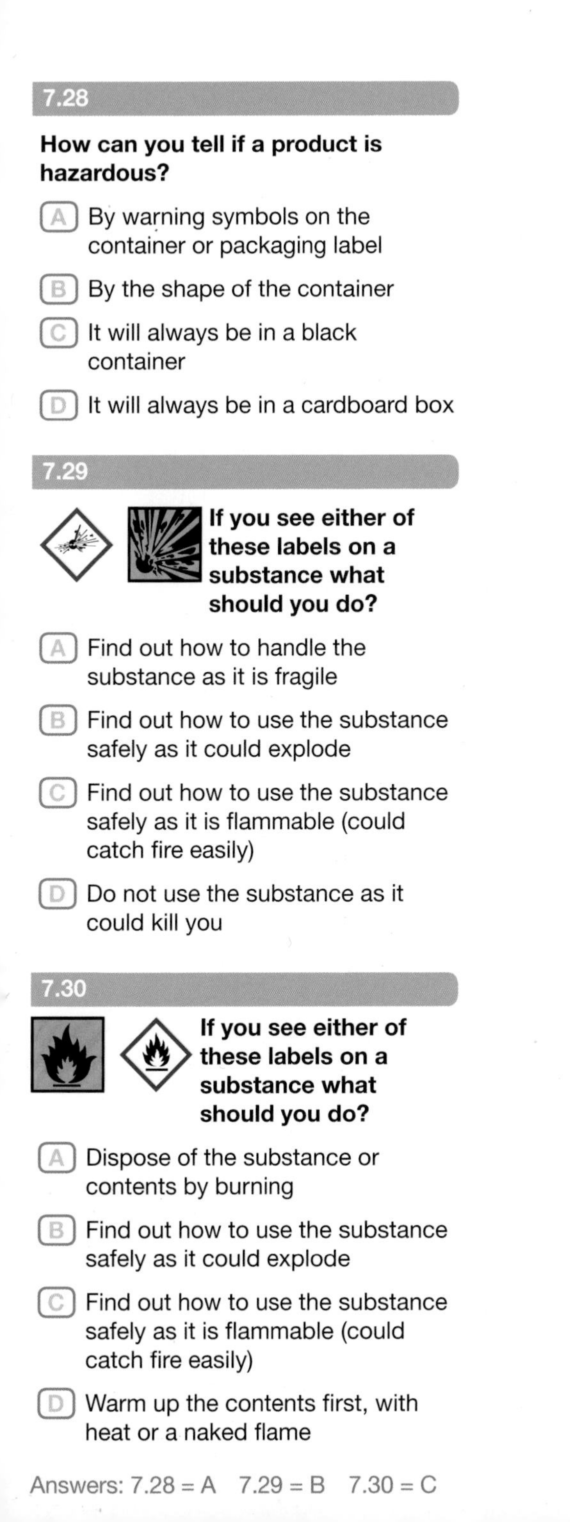

7.28

How can you tell if a product is hazardous?

A) By warning symbols on the container or packaging label

B) By the shape of the container

C) It will always be in a black container

D) It will always be in a cardboard box

7.29

If you see either of these labels on a substance what should you do?

A) Find out how to handle the substance as it is fragile

B) Find out how to use the substance safely as it could explode

C) Find out how to use the substance safely as it is flammable (could catch fire easily)

D) Do not use the substance as it could kill you

7.30

If you see either of these labels on a substance what should you do?

A) Dispose of the substance or contents by burning

B) Find out how to use the substance safely as it could explode

C) Find out how to use the substance safely as it is flammable (could catch fire easily)

D) Warm up the contents first, with heat or a naked flame

Answers: 7.28 = A 7.29 = B 7.30 = C

C
General safety

8.1

While on site you observe that there is a risk of materials flying at speed into site workers' eyes. What should they be wearing in order to protect themselves?

A Impact-resistant goggles or full face shield

B Welding goggles

C Reading glasses or sunglasses

D Light eye protection

8.2

When should eye protection be worn?

A On very bright, sunny days

B If there is a risk of eye injury and if it is the site rules

C When it has been included in the bill of quantities

D Only for work with chemicals

8.3

What type of eye protection do you need to wear if you are using a cartridge-operated tool or compressed gas tool (nail gun)?

A Light eye protection or safety glasses

B Normal prescription glasses or sunglasses

C Impact-rated goggles

D None – they aren't needed as there is a minimal risk of injury

8.4

When should you wear safety footwear on site?

A Only when working at ground level or outside

B Until the site starts to look finished

C All the time

D When you are working all day on site and not just visiting

8.5

What features should you be looking for when obtaining safety footwear for a site visit?

A They must be black with a good sole pattern

B They only need a protective toecap

C They must have a protective toecap and mid-sole

D They must be smooth-soled to prevent the transfer of contaminated materials

8.6

Do all types of glove protect hands against chemicals?

A Yes, all gloves are made to the same standard

B Only if site workers put barrier cream on their hands as well

C No, different types of glove protect against different types of hazard

D Only if site workers cover their gloves with barrier cream

Answers: 8.1 = A 8.2 = B 8.3 = C 8.4 = C 8.5 = C 8.6 = C

8.7

Look at these statements about anti-vibration gloves. Which one is true?

- A They might cut out some hand-arm vibration
- B They cut out most hand-arm vibration
- C They cut out all hand-arm vibration
- D They give the most protection if they are worn over other gloves

8.8

To get the maximum protection from your safety helmet you should wear it:

- A back to front
- B pushed back on your head
- C square on your head
- D pulled forward

8.9

You must wear head protection on site at all times unless you are:

- A working on a project that is at the finishing stages
- B working where there are no hazards above you
- C in a safe area, like the site office
- D working in very hot weather

8.10

If you drop your safety helmet from height on to a hard surface, you should:

- A have any cracks repaired then carry on wearing it
- B make sure there are no cracks then carry on wearing it
- C work without a safety helmet until you can get a new one
- D stop work and get a new safety helmet

8.11

You must wear hi-vis clothing:

- A when the need is identified in the contractor's or your employer's site rules
- B only if you are inspecting deep excavations or tunnels
- C during normal daylight hours only
- D only if you are working alongside moving plant

8.12

If your personal protective equipment (PPE) gets damaged you should:

- A throw it away and work without it
- B stop what you are doing until it is replaced
- C carry on wearing it but work more quickly
- D try to repair it

Answers: 8.7 = A 8.8 = C 8.9 = C 8.10 = D 8.11 = A 8.12 = B

8.13

You are about to enter an active work area on site. How will you know if you need any extra personal protective equipment (PPE)?

A By looking at your employer's health and safety policy

B You will just be expected to know

C From the risk assessment or method statement

D Others around you will be wearing more than the minimum PPE required

8.14

Who has the legal duty to ensure that workers are provided with any personal protective equipment (PPE) they need, including the means to maintain it?

A Their employer

B The workers who need it

C The client for the project

D The person whose design created the need for the use of PPE

8.15

You have to work outdoors in bad weather. Your employer should supply you with waterproof clothing because:

A it will have the company name and logo on it

B you need protecting from the weather and are less likely to get muscle strains if you are warm and dry

C you are less likely to catch Weil's disease (leptospirosis) if you are warm and dry

D your supervisor will be able to see you more clearly in the rain

8.16

Look at these statements about personal protective equipment (PPE). Which one is NOT true? Workers must:

A pay for any damage or loss

B store it correctly when they are not using it

C report any damage or loss to their manager

D use it as instructed

Answers: 8.13 = C 8.14 = A 8.15 = B 8.16 = A

8.17

Your employer must supply personal protective equipment (PPE):

A twice a year

B if workers pay for it

C if it is in the contract

D if it is needed to provide protection

8.18

Do workers have to pay for any personal protective equipment (PPE) they need?

A Yes, they must pay for all of it

B Only to replace lost or damaged PPE

C Yes, but they only have to pay half the cost

D No, the employer must pay for it

Answers: 8.17 = D 8.18 = D

9.1

Untidy leads and extension cables are responsible for many trips and lost work time injuries. What TWO things should you do to help?

A Run cables and leads above head height and over the top of doorways and walkways rather than across the floor

B Tie any excess cables and leads up into the smallest coil possible

C Keep cables and leads close to the wall and not in the middle of the floor or walkway

D Make sure your cables go where you want them to and not worry about others

E Unplug the nearest safety lighting and use these sockets instead

9.2

What is the best way to protect an extension cable while you work, as well as minimising trip hazards?

A Run the cable above head height

B Run the cable by the shortest route

C Cover the cable with yellow tape

D Cover the cable with pieces of wood

9.3

If the guard is missing from a power tool you should:

A try to make another guard

B use the tool but try to work quickly

C not use the tool until a proper guard has been fitted

D use the tool but work carefully and slowly

9.4

If you need to use a power tool with a rotating blade, you should:

A remove the guard so that you can clearly see the blade

B adjust the guard to expose just enough blade to let you do the job

C remove the guard but wear leather gloves to protect your hands

D adjust the guard to expose the maximum amount of blade

9.5

Most cutting and grinding machines have guards. What are the TWO main functions of the guard?

A To stop materials getting onto the blade or wheel

B To give the operator a firm handhold

C To balance the machine

D To stop fragments flying into the air

E To stop the operator coming into contact with the blade or wheel

Answers: 9.1 = A, C 9.2 = A 11.3 = C 9.4 = B 9.5 = D, E

9.6

Someone near you is using a rotating laser level. What, if any, is the health hazard likely to affect you?

A Skin cancer

B None – if used correctly they are safe

C Gradual blindness

D Burning of the skin, similar to sunburn

9.7

It is dangerous to run an abrasive wheel faster than its recommended top speed. Why?

A The wheel will get clogged and stop

B The motor could burst into flames

C The wheel could shatter and burst into many pieces

D The safety guard cannot be used

9.8

It is safe to work close to an overhead power line if:

A you do not touch the line

B you use a wooden ladder

C there is a clear indication that the power is switched off

D it is not raining

9.9

You are inspecting a site where there are overhead electric cables. What arrangements should a contractor normally have in place to alert those on site to the presence of the cables? Give TWO answers.

A Scaffolding fan

B Warning signs

C Gates

D Barriers and height restriction goalposts

E Traffic lights

9.10

Someone near you is using a disc-cutter to cut concrete blocks. What THREE immediate hazards are likely to affect you?

A Flying fragments

B Dermatitis

C Harmful dust in the air

D High noise levels

E Vibration white finger

9.11

When do you need to check tools and equipment for damage?

A Each time before use

B Every day

C Once a week

D At least once a year

Answers: 9.6 = B 9.7 = C 9.8 = C 9.9 = B, D 9.10 = A, C, D 9.11 = A

9.12

What are the TWO main areas of visual inspections you should carry out before each use of a power tool?

A Check the carry case isn't broken

B Check the power lead, plug and casing are in good condition

C Check the manufacturer's label hasn't come off

D Check switches, triggers and guards are adjusted and work correctly

E Check if there is an upgraded model available

9.13

A RCD (residual current device) must be used in conjunction with 230 volt electrical equipment because it:

A lowers the voltage

B quickly cuts off the power if there is a fault

C makes the tool run at a safe speed

D saves energy and lowers costs

9.14

How could a site worker check if the RCD (residual current device) through which a 230 volt hand tool is connected to the supply is working correctly?

A Switch the tool on and off

B Press the test button on the RCD unit

C Switch the power on and off

D Run the tool at top speed to see if it cuts out

9.15

Which TWO of the following would you expect to find on a PAT test label?

A The date when the next safety check is due

B When the equipment was made

C Who tested the equipment before it left the factory

D Its earth-loop impedance

E The date when the equipment was last tested as being safe to use (pass)

9.16

This warning sign means:

A risk of electrocution

B risk of radiation

C electrical appliance

D risk of lightning

9.17

The colour of a 110 volt power cable and connector should be:

A black

B red

C blue

D yellow

Answers: 9.12 = B, D   9.13 = B   9.14 = B   9.15 = A, E   9.16 = A   9.17 = D

9.18

Why do building sites use a 110 volt electricity supply instead of the usual 230 volt domestic supply?

A It is cheaper

B It is less likely to kill people

C It moves faster along the cables

D It is safer for the environment

9.19

What is the significance of a yellow plug and a yellow supply cable fitted to an electrical hand tool?

A The tool runs off a 110 volt supply

B The tool is waterproof and can be used outdoors in wet conditions

C The tool runs off a 240 volt supply and should not be used on site

D The tool has been PAT tested within the past 12 months

9.20

On building sites the recommended safe voltage for electrical equipment is:

A 12 volts

B 24 volts

C 110 volts

D 230 volts

9.21

In the colour coding of electrical power supplies on site, what voltage does a blue plug represent?

A 50 volts

B 110 volts

C 240 volts

D 415 volts

9.22

On the site electrical distribution system, which colour plug indicates a 415 volt supply?

A Yellow

B Blue

C Black

D Red

Answers: 9.18 = B 9.19 = A 9.20 = C 9.21 = C 9.22 = D

10.1

A crane has to do a difficult lift. The signaller asks you to help, but you are not trained in plant signals. What should you do?

A Politely refuse because you don't know how to signal

B Start giving signals to the crane driver

C Only help if the signaller really can't manage alone

D Ask the signaller to show you what signals to use

10.2

A truck has to tip materials into a trench. Who should give signals to the truck driver?

A Anyone who is wearing a hi-vis coat

B Someone standing in the trench

C Someone who knows the signals

D Only the person who is trained and appointed for the job

10.3

These signs tell you that a substance can be:

A harmful

B toxic

C corrosive

D dangerous to the environment

10.4

What does this sign mean?

A Assemble here in the event of a fire

B Fire extinguishers and fire-fighting equipment kept here

C Parking reserved for emergency service vehicles

D Do not store flammable materials here

10.5

What does this sign mean?

A Fire alarm call point

B Hot surface, do not touch

C Wear flameproof hand protection

D Emergency light switch

10.6

What does this sign mean?

A Press here to sound the fire alarm

B Fire hose reel located here

C Turn key to open fire door

D Do not use if there is a fire

Answers: 10.1 = A 10.2 = D 10.3 = D 10.4 = B 10.5 = A 10.6 = B

10.7

What does this sign mean?

- A Safety boots or safety shoes must be worn
- B Wellington boots must be worn
- C Be aware of slip and trip hazards
- D No dirty footwear past this point

10.8

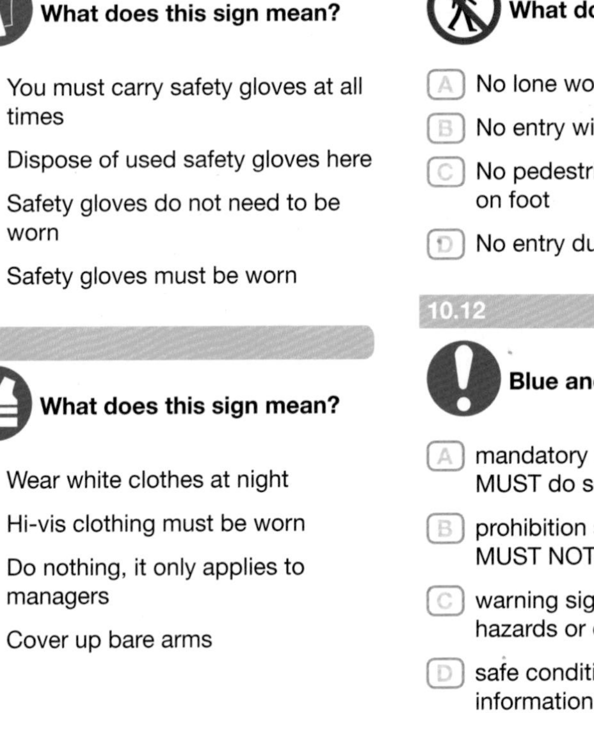

What does this sign mean?

- A You must carry safety gloves at all times
- B Dispose of used safety gloves here
- C Safety gloves do not need to be worn
- D Safety gloves must be worn

10.9

What does this sign mean?

- A Wear white clothes at night
- B Hi-vis clothing must be worn
- C Do nothing, it only applies to managers
- D Cover up bare arms

10.10

What does this sign mean?

- A Safety glasses cleaning station
- B Warning, bright lights or lasers
- C Caution, poor lighting
- D You must wear safety eye protection

10.11

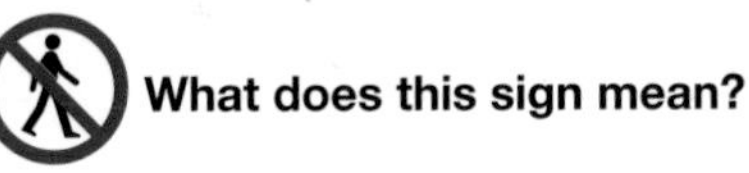

What does this sign mean?

- A No lone working
- B No entry without a hard hat
- C No pedestrians or entry for people on foot
- D No entry during the day

10.12

Blue and white signs are:

- A mandatory signs – meaning you MUST do something
- B prohibition signs – meaning you MUST NOT do something
- C warning signs – alerting you of hazards or danger
- D safe condition signs – giving you information

Answers: 10.7 = A 10.8 = D 10.9 = B 10.10 = D 10.11 = C 10.12 = A

10.13

Round red and white signs with a diagonal line are:

- A mandatory signs – meaning you MUST do something
- B prohibition signs – meaning you MUST NOT do something
- C warning signs – alerting you of hazards or danger
- D safe condition signs – giving you information

10.14

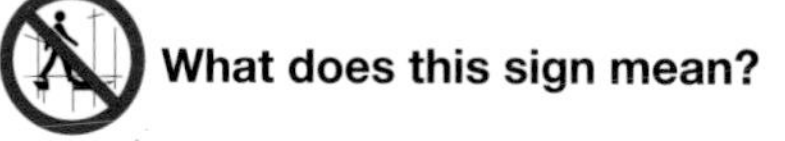

What does this sign mean?

- A Do not jump across any gaps in the scaffold
- B Do not work on the first lift of the scaffold
- C Do not access the scaffold because it is incomplete or not safe
- D Do not walk under the scaffold

10.15

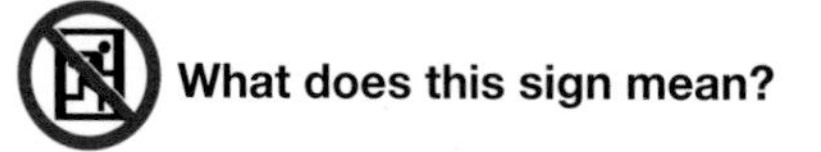

What does this sign mean?

- A No running allowed
- B There is no escape route
- C This is a fire door
- D Fire escape route

10.16

What does this sign mean?

- A Wear hearing protection if you want to
- B You must wear hearing protection
- C No personal stereos or MP3 players
- D Caution, noisy machinery

10.17

Emergency and safe condition signs, such as fire exit and first aid, are coloured:

- A blue and white
- B red and white
- C green and white
- D red and yellow

10.18

Green and white signs are:

- A mandatory signs – meaning you MUST do something
- B prohibition signs – meaning you MUST NOT do something
- C warning signs – alerting you of hazards or danger
- D safe condition signs – giving you information

Answers: 10.13 = B 10.14 = C 10.15 = B 10.16 = B 10.17 = C 10.18 = D

10.19

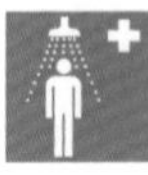

What does this sign mean?

- A Toilets and shower facilities
- B Drying area for wet weather clothes
- C Emergency first-aid shower
- D Fire sprinklers above

10.20

What does this sign mean?

- A Dispose of substance or contents by burning
- B Warning – substance or contents are flammable (can catch fire easily)
- C Warning – substance or contents could explode
- D Warning – substance or contents are harmful

10.21

What does this sign mean?

- A Radioactive area
- B Warning – explosive substance
- C Flashing lights ahead
- D Warning – laser beams

10.22

Yellow and black signs are:

- A mandatory signs – meaning you MUST do something
- B prohibition signs – meaning you MUST NOT do something
- C warning signs – alerting you of hazards or danger
- D safe condition signs – giving you information

10.23

What does this sign mean?

- A Plant operators wanted
- B Industrial vehicles are moving about
- C Manual handling is not allowed
- D Storage area

Answers: 10.19 = C 10.20 = B 10.21 = D 10.22 = C 10.23 = B

11.1

Which of the following, taken on their own, provides the best solution for reducing risks related to site transport and access?

- A Provide all site staff with information detailing the site layout and designated traffic routes
- B A one-way traffic system complete with segregated pedestrian routes
- C Ensure that there are adequate signs directing traffic to various parts of the site
- D Reduce the need for some vehicle movements on site by requiring more materials to be offloaded manually outside the site

11.2

Which TWO of the following conditions would you expect a site manager to apply in order to manage the risk of site staff operating plant?

- A The plant operative must be trained and competent
- B The plant operative must be authorised
- C The plant operative must be over 21 years old
- D The plant operative must hold a full driving licence
- E The plant operative must be under 65 years old

11.3

If you see a dumper being driven too fast, you should:

- A keep out of its way and report the matter to the site manager
- B try to catch the dumper and speak to the driver
- C report the matter to the Health and Safety Executive (HSE)
- D do nothing, dumpers are allowed to go above the site speed limit

11.4

While observing an excavator digging a trial pit you notice that liquid is dripping and forming a small pool under the back of the machine. What could this mean?

- A It is normal for fluids to vent after the machine stops
- B The machine is hot so the diesel has expanded and overflowed
- C Someone put too much diesel into the machine before it started work
- D The machine may have an hydraulic fluid leak and could be unsafe

11.5

How should you be told about the site traffic rules?

- A During site induction
- B By a Health and Safety Executive (HSE) inspector
- C By a note on the site notice or hazard board
- D By the plant operators

Answers: 11.1 = B 11.2 = A, B 11.3 = A 11.4 = D 11.5 = A

11.6

A mobile plant operator can let people ride in their machine:

- A if they have a long way to go
- B if it is raining
- C if it is designed to carry passengers
- D at any time

11.7

While visiting site you notice a build-up of diesel fumes in the area of the site that you are visiting. What is the correct course of action that you should take?

- A Turn off the piece of plant that is creating the fumes
- B Quickly inform the site manager of this hazardous situation
- C Carry out the visit quickly to minimise exposure
- D Move out of the affected area at regular intervals to get fresh air

11.8

Which of the following represents good site management on the public road approaching a site?

- A A place where drivers can park delivery lorries off the road
- B Items of plant parked to free up space on site
- C 'Apologies for any inconvenience caused' signs
- D Contractors asked to park half on the footpath and half on the road so the site entrance can be seen more easily

11.9

You see a mobile crane lifting a load. The load is about to hit something. What should you do?

- A Warn the site manager
- B Warn the person supervising the lift
- C Warn the crane driver
- D Do nothing and assume everything is under control

11.10

You think a load is about to fall from a moving fork-lift truck. What should you do?

- A Keep clear but try to warn the driver and others in the area
- B Run alongside the machine and try to hold on to the load
- C Run and tell the site manager
- D Sound the nearest fire alarm bell

11.11

The correct procedure for using a tower crane to offload a lorry is for:

- A lorry drivers to sling the load before the trained slinger/signaller arrives
- B anyone to sling the load, providing it will not pass over people when on the crane
- C a trained slinger/signaller to carry out the offloading operation
- D the crane driver to instruct an operative to sling the load

Answers: 11.6 = C 11.7 = B 11.8 = A 11.9 = B 11.10 = A 11.11 = C

11.12

All lifting equipment and accessories should be:

- A brightly coloured, inspected and clearly signed
- B regularly maintained, clean and tidy
- C logged, inspected, thoroughly examined and marked
- D strong enough for the load and always fitted with outriggers

11.13

You need to walk past someone using a mobile crane. You should:

- A anticipate what the crane operator will do next and then pass
- B try to catch the attention of the crane operator first
- C walk past but only if you are wearing Class 2 or Class 3 hi-vis clothing
- D take another route so that you stay clear of the crane

11.14

You need to walk past a 360° mobile crane. The crane is operating near a wall. What is the main danger?

- A You may put the crane driver off if he suddenly sees you
- B You could be crushed if you walk between the crane and the wall
- C The crane's diesel exhaust fumes could build up near the wall and become a hazard to you
- D Noise levels may increase above safe levels as they will echo off the wall

11.15

You are walking across the site. A large mobile crane reverses across your path. What should you do?

- A Help the driver to reverse
- B Start to run so that you can pass behind the reversing crane
- C Pass close to the front of the crane
- D Wait or find another way around the crane

11.16

When is site transport allowed to drive along a pedestrian route?

- A During meal breaks
- B If it is the shortest route
- C Only if necessary and if all pedestrians are excluded
- D Only if the vehicle has a flashing yellow light

11.17

How would you expect a well-organised site to keep pedestrians away from traffic routes?

- A The site manager will direct all pedestrians away from traffic routes
- B The traffic routes will be shown on the site notice or hazard board
- C There will be physical barriers between traffic and pedestrian routes
- D The plant operators will be given strict instructions on which route they must take

Answers: 11.12 = C 11.13 = D 11.14 = B 11.15 = D 11.16 = C 11.17 = C

11.18

A site vehicle is most likely to injure pedestrians when it is:

A reversing

B lifting materials onto scaffolds

C tipping into an excavation

D digging out footings

11.19

You must not walk behind a lorry when it is reversing because:

A most lorries are not fitted with mirrors

B the driver is unlikely to know you are there

C the driver may think you are the signaller

D you could be overcome by exhaust fumes

11.20

The easiest way to where you want to be on site is through a contractor's vehicle compound. Which route should you take?

A Around the compound if vehicles are moving

B Straight through the compound if no vehicles appear to be moving

C Around the compound every time

D Through the compound but staying close to the edge away from vehicles

11.21

A fork-lift truck is blocking the way to where you want to go on site. It is lifting materials on to a scaffold. What should you do?

A Only walk under the raised load if you are wearing a safety helmet

B Catch the driver's attention and then walk under the raised load

C Start to run so that you are not under the load for very long

D Wait or go around, but never walk under a raised load

11.22

Which of the following is the most effective way of preventing pedestrians being struck by site vehicles?

A All vehicles must switch on their flashing amber beacon

B Separate access gates and routes for pedestrians and vehicles

C Hi-vis vests being worn when pedestrians walk up the site road

D A wide site road with a good quality surface

11.23

Of the following, which is the best risk control measure with regard to site vehicles reversing?

A Setting a speed limit on site

B Vehicles fitted with reversing bleepers

C A signaller to reverse all vehicles, especially on and off site

D All vehicles fitted with CCTV to help them reverse

Answers: 10.18 = A 11.19 = B 11.20 = C 11.21 = D 11.22 = B 11.23 = C

11.24

When you walk across the site, what is the best way to avoid an accident with mobile plant?

A Keep to the designated pedestrian routes

B Keep to the routes everyone else is taking

C Get the attention of the driver before you get too close

D Wear hi-vis clothing

11.25

Which of these would you NOT expect to see if site transport is well organised?

A Speed limits

B Barriers to keep pedestrians away from mobile plant and vehicles

C Pedestrians and mobile plant using the same routes

D One-way systems

11.26

Before allowing a lifting operation to be carried out, the contractor must ensure that the sequence of operations to enable a lift to be carried out safely is confirmed in:

A verbal instructions

B a lift plan or method statement

C a tool box talk

D a risk assessment

Answers: 11.24 = A 11.25 = C 11.26 = B

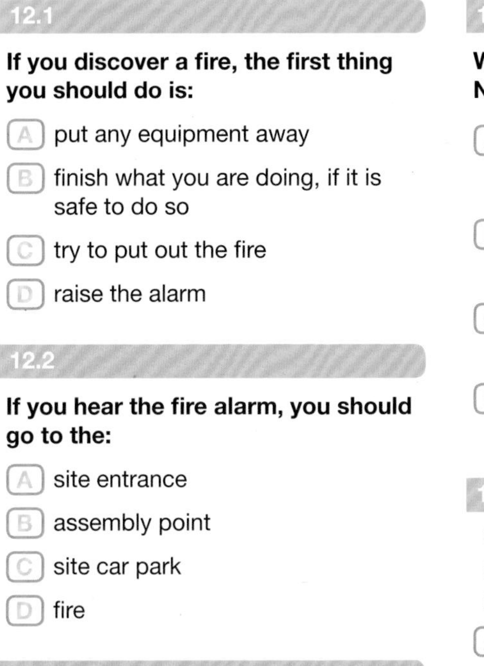

12.1

If you discover a fire, the first thing you should do is:

- A put any equipment away
- B finish what you are doing, if it is safe to do so
- C try to put out the fire
- D raise the alarm

12.2

If you hear the fire alarm, you should go to the:

- A site entrance
- B assembly point
- C site car park
- D fire

12.3

When using a carbon dioxide (CO2) fire extinguisher, you must not touch the nozzle because it gets:

- A very cold
- B very hot
- C sticky
- D very heavy

12.4

Which TWO extinguishers should NOT be used on electrical fires?

- A  Dry powder (Blue colour band)
- B  Foam (Cream colour band)
- C  Water (Red colour band)
- D Carbon dioxide (Black colour band)

12.5

What type of fire extinguisher should NOT be provided where petrol or diesel is being stored?

- A Foam
- B Water
- C Dry powder
- D Carbon dioxide

12.6

A WATER fire extinguisher, identified by a red band, should ONLY be used on what type of fire?

- A Wood, paper, textile and solid material fires
- B Flammable liquids (fuel, oil, varnish, paints, etc.)
- C Electrical fires
- D Metal and molten metal

Answers: 12.1 = D 12.2 = B 12.3 = A 12.4 = B, C 12.5 = B 12.6 = A

12.7

A DRY POWDER fire extinguisher, identified by a blue band, could be used on all types of fire but is BEST suited to what TWO types of fire?

- A Wood, paper, textile and solid material fires
- B Flammable liquids (fuel, oil, varnish, paints, etc.)
- C Flammable gas (LPG, propane, etc.)
- D Metal and molten metal
- E Electrical fires

12.8

A FOAM extinguisher, identified by a cream band, should NOT be used on what TWO types of fire?

- A Wood, paper, textile and solid material fires
- B Flammable liquids (fuel, oil, varnish, paints, etc.)
- C Electrical fires
- D Metal and molten metal
- E Flammable gas (LPG, propane, etc.)

12.9

A CARBON DIOXIDE (CO2) extinguisher, identified by a black band, should NOT be used on what type of fire?

- A Wood, paper, textile and solid material fires
- B Flammable liquids (fuel, oil, varnish, paints, etc.)
- C Electrical fires
- D Metal and molten metal

12.10

If you see 'frost' around the valve on a liquefied petroleum gas (LPG) cylinder, it means:

- A the cylinder is nearly empty
- B the cylinder is full
- C the valve is leaking
- D you must lay the cylinder on its side

12.11

All fires need heat, fuel and oxygen. Knowing this, explain how a water extinguisher puts out a wood fire.

- A By removing the oxygen
- B By smothering the flames
- C By removing the fuel
- D By cooling the fuel

Answers: 12.7 = B, C 12.8 = C, D 12.9 = D 12.10 = C 12.11 = D

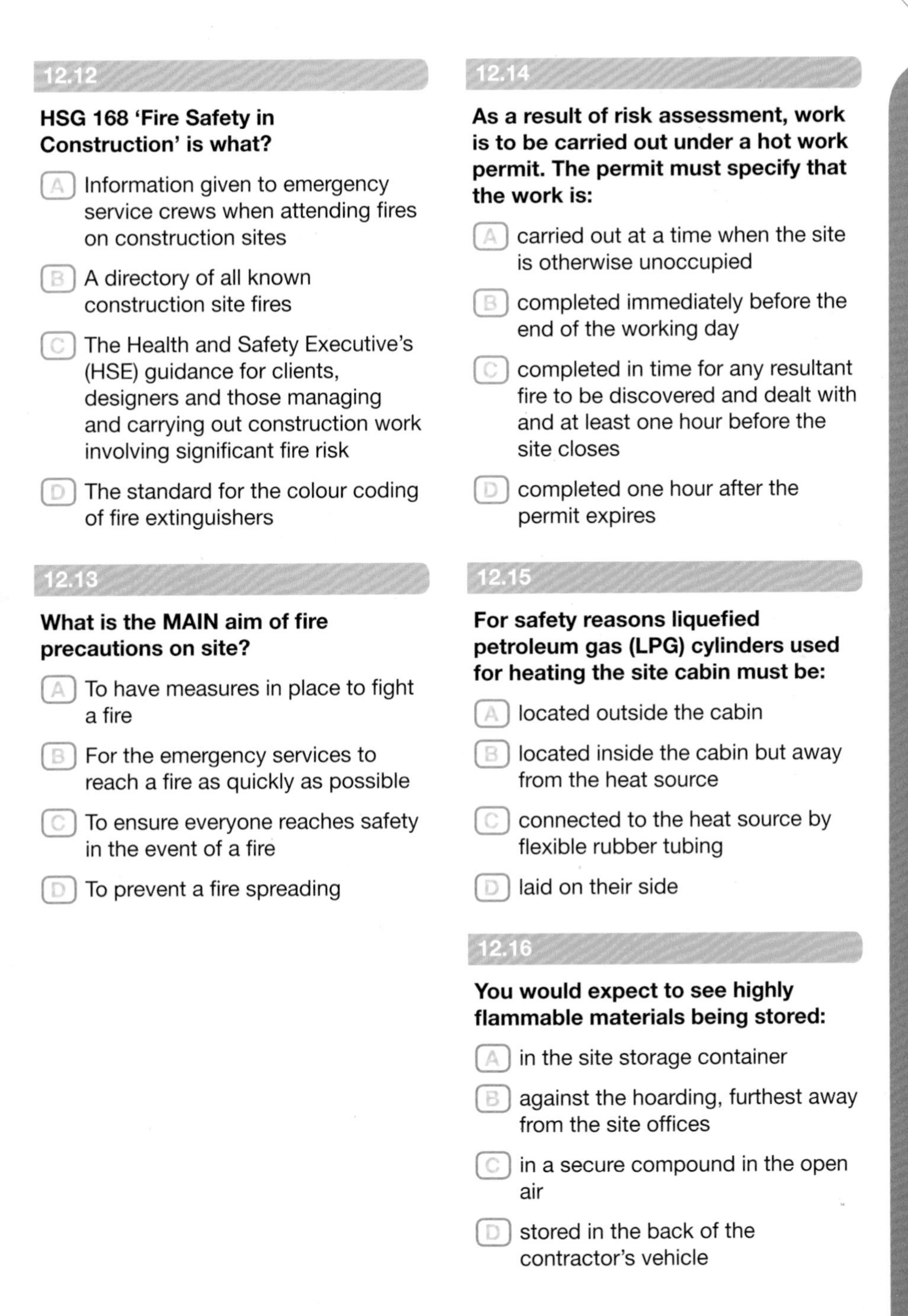

12.12

HSG 168 'Fire Safety in Construction' is what?

A Information given to emergency service crews when attending fires on construction sites

B A directory of all known construction site fires

C The Health and Safety Executive's (HSE) guidance for clients, designers and those managing and carrying out construction work involving significant fire risk

D The standard for the colour coding of fire extinguishers

12.13

What is the MAIN aim of fire precautions on site?

A To have measures in place to fight a fire

B For the emergency services to reach a fire as quickly as possible

C To ensure everyone reaches safety in the event of a fire

D To prevent a fire spreading

12.14

As a result of risk assessment, work is to be carried out under a hot work permit. The permit must specify that the work is:

A carried out at a time when the site is otherwise unoccupied

B completed immediately before the end of the working day

C completed in time for any resultant fire to be discovered and dealt with and at least one hour before the site closes

D completed one hour after the permit expires

12.15

For safety reasons liquefied petroleum gas (LPG) cylinders used for heating the site cabin must be:

A located outside the cabin

B located inside the cabin but away from the heat source

C connected to the heat source by flexible rubber tubing

D laid on their side

12.16

You would expect to see highly flammable materials being stored:

A in the site storage container

B against the hoarding, furthest away from the site offices

C in a secure compound in the open air

D stored in the back of the contractor's vehicle

Answers: 12.12 = C 12.13 = C 12.14 = C 12.15 = A 12.16 = C

12.17

Acetylene and oxygen cylinders that are not in use are being stored together outside a meeting room. In the interests of everyone's safety they should be:

A kept on a bottle trolley together

B stored separately and away from site accommodation

C laid down so that they cannot fall over and damage the valves

D stored together but away from site accommodation

12.18

Work is taking place in a corridor that is a fire escape route. The contractor must ensure that:

A tools, equipment and materials do not block the route

B all doors into the corridor are locked

C only spark-proof tools are used

D all fire escape signs are removed before work starts

12.19

Major fires are rare on site, but when they happen fire and smoke can rapidly spread. What should be in place on all construction sites to ensure precautions required are adequately assessed?

A A fire plan/risk assessment carried out at the start of the project

B Hot work permits

C A construction phase health and safety plan

D A fire plan/risk assessment regularly updated to reflect current site conditions

12.20

What are TWO common fire risks on construction sites?

A 230 volt power tools

B Poor housekeeping and build up of waste

C Timber racks

D Uncontrolled hot works

E 110 volt extension reels

Answers: 12.17 = B 12.18 = A 2.19 = D 12.20 = B, D

D

High risk activities

13.1

Under the regulations for manual handling, all workers must:

- A only exceed the weights identified in the risk assessment if they know they are capable of lifting them
- B make a list of all the heavy things they have to carry
- C lift any size of load they feel comfortable with
- D follow the requirements of their employer's safe systems of work

13.2

If manual lifting activities are part of a task, what must the employer do?

- A Make sure the lifting operations are supervised
- B Carry out a risk assessment of the task
- C Nothing, as it is part of some work operations to lift loads
- D Watch while the load is lifted

13.3

You are in charge of a gang of workers and about to start on a new site. What can you do to help minimise manual handling?

- A Leave it to the workforce to sort out – they always find a way
- B Hire in extra labour to carry the materials and equipment
- C Assess and agree with the site manager how the materials and equipment can be distributed close to the workplace
- D Make sure the risk assessment has the correct site name on it

13.4

Your workforce are obviously lifting more than what the risk assessment states. What should you do?

- A Find out how much they are lifting and change the risk assessment to this weight
- B Tell them they shouldn't, but let them carry on. At least you have warned them
- C Let them carry on, they are doing the work quicker and seem OK
- D Stop them, find out why, agree a solution and amend the risk assessment

Answers: 13.1 = D 13.2 = B 13.3 = C 13.4 = D

13.5

Who should decide what weight is safe for you to lift as part of a safe system of work?

(A) You

(B) Your supervisor/employer

(C) You and your supervisor/employer

(D) The Health and Safety Executive (HSE)

13.6

You are using a wheelbarrow to move a heavy load. Is this manual handling?

(A) No, because the wheelbarrow is carrying the load

(B) Only if the load slips off the wheelbarrow

(C) Yes, you are still manually handling the load

(D) Only if the wheelbarrow has a flat tyre

13.7

Your new job involves some manual handling of survey equipment in and out of vehicles. An old injury means that you have a weak back. What should you do?

(A) If you take care there is no need to inform your manager

(B) Tell your manager that lifting might be a problem

(C) Try some lifting then tell your manager about your back

(D) Only tell your manager about your back if it becomes a problem

13.8

You have been asked to move a load that might be too heavy for you. You cannot divide it into smaller parts and there is no-one to help you. What should you do?

(A) Do not move the load until you have found a safe method

(B) Ask someone to get a fork-lift truck, even though you suspect they can't drive one

(C) You know how to lift, so try to lift it using the correct lifting methods

(D) Get on and lift and move the load quickly as this helps avoid injury

13.9

You need to lift a load from the floor. You should stand with your:

(A) feet together, legs straight, back bent

(B) feet together, knees bent, in a deep squatting position

(C) feet slightly apart, one leg slightly forward, knees flexed

(D) feet wide apart, legs straight, back bent

Answers: 13.5 = C 13.6 = C 13.7 = B 13.8 = A 13.9 = C

13.10

If you have to twist or turn your body when you lift and place a load, it means:

A the weight you can lift safely is LESS than usual

B the weight you can lift safely is MORE than usual

C nothing, you can lift the SAME weight as usual

D you MUST wear a back brace

13.11

Someone has to move a load while they are sitting, not standing. How much can they move safely?

A Less than usual

B The usual amount

C Twice the usual amount

D Three times the usual amount

Answers: 13.10 = A   13.11 = A

14.1

If someone is wearing a harness while working at height, what else must be done?

- A Provide an extra harness in case theirs breaks
- B Nothing else, wearing a harness is good enough
- C Have a rescue plan in place to retrieve them quickly if they fall
- D Have a second person warn them if they get too close to the edge

14.2

What is the main danger of leaving someone who has fallen suspended in a harness for too long?

- A The anchorage point may fail
- B They may try to climb back up the structure and fall again
- C They may suffer severe trauma or even death
- D It is a distraction for other workers

14.3

When is it most appropriate to use a safety harness and lanyard for working at height?

- A Only when the roof has a steep pitch
- B Only when crossing a flat roof with clear rooflights
- C Only when all other options for fall prevention have been ruled out
- D Only when materials are stored at height

14.4

If a fall-restraint lanyard has damaged stitching, the user should:

- A use the lanyard if the damaged stitching is less than two inches long
- B get a replacement lanyard
- C do not use the damaged lanyard and work without one
- D use the lanyard if the damaged stitching is less than six inches long

14.5

In order to carry out a structural inspection you need to wear a full body harness. You have never used one before. What must happen before you start work?

- A Your employer must provide you with information, competent advice and training
- B Ask someone wearing a similar harness to show you what to do
- C Try to work it out for yourself
- D Read the instruction book and follow any advice that it contains

Answers: 14.1 = C 14.2 = C 14.3 = C 14.4 = B 14.5 = A

14.6

A design feature of some airbags used for fall arrest is a controlled leak rate. If you are using these, the inflation pump must:

- A be electrically powered
- B be switched off from time to time to avoid over-inflation
-  run all the time while work is carried out at height
- D be switched off when the airbags are full

14.7

Why is it dangerous to use inflatable airbags for fall arrest that are too big for the area to be protected?

- A They will exert a sideways pressure on anything that is containing them
- B The pressure in the bags will cause them to burst
- C The inflation pump will become overloaded
- D They will not fully inflate

14.8

On a working platform, the maximum permitted gap between the guard-rails is:

- A 350 mm
- B 470 mm
- C 490 mm
- D 510 mm

14.9

Under the requirements of the Work at Height Regulations, the minimum width of a working platform must be:

- A two scaffold boards wide
- B three scaffold boards wide
- C four scaffold boards wide
- D suitable and sufficient for the job in hand

14.10

On a working platform, the minimum height of the main guard-rail must be:

- A 750 mm
- B 850 mm
- C 950 mm
- D 1,050 mm

14.11

The Beaufort Scale is important when planning any external work at height because it measures:

- A air temperature
- B the load-bearing capacity of a flat roof
- C wind speed
- D the load-bearing capacity of a scaffold

Answers: 14.6 = C   14.7 = A   14.8 = B   14.9 = D   14.10 = C   14.11 = C

14.12

What is the most effective method to prevent workers falling from height while carrying out construction and maintenance work?

A Leave the decisions on how to work at height to the principal contractor

B Ensure details of risky operations are included in the construction phase health and safety plan

C Educate the workforce to be more careful while working at height

D Ensure that design and construction solutions eliminate the need for working at height

14.13

What is the main reason for using a safety net or other soft-landing system rather than a personal fall-arrest system?

A Soft-landing systems are cheaper to use and do not need inspecting

B It is always easy to rescue workers who fall into a soft-landing system

C Specialist knowledge is not required to install soft-landing systems

D Soft-landing systems are 'collective' fall arrest measures

14.14

Edge protection must be designed to:

A allow persons to work both sides

B secure tools and materials close to the edge

C warn people where the edge of the roof is

D prevent people and materials falling

14.15

In law you are working at height when you could fall from:

A the first lift of a scaffold or higher

B 2 m above the ground or higher

C any height that would cause an injury if you fell

D 3 m above the ground or higher

14.16

Following the principles of prevention, which of the following is to be regarded as the last resort for someone's safety when working at height?

A Safety harness and fall arrest lanyard

B Safety netting or airbags

C Mobile elevating work platform (MEWP)

D Access tower scaffold

Answers: 14.12 = D 14.13 = D 14.14 = D 14.15 = C 14.16 = A

14.17

What is the best way to make sure that a ladder is secure and won't slip?

- (A) Ensure that it is tied at the top
- (B) Ask someone to stand with their foot on the bottom rung
- (C) Tie it at the bottom
- (D) Ask for the bottom of the ladder to be wedged with blocks of wood

14.18

To ensure the safety of people who have to gain access to a place of work at height, ladders are:

- (A) always acceptable for work below 2 m
- (B) alright to use if it gets the job done more quickly
- (C) generally the least favoured option in the hierarchy of risk
- (D) now banned on all sites

14.19

How far should a ladder extend above the stepping-off point if there is no alternative, firm handhold?

- (A) Two rungs
- (B) Three rungs
- (C) Five rungs or one metre
- (D) Half a metre

14.20

When using ladders for access, what is the maximum vertical distance between landings?

- (A) 5 m
- (B) There is no maximum
- (C) 9 m
- (D) 30 m

14.21

A 'Class 3' ladder is:

- (A) for domestic use only and must not be used at work
- (B) of industrial quality and can be used at work
- (C) a ladder that has been made to a European Standard
- (D) made of insulating material and can be used near to overhead cables

14.22

When using a ladder what should the slope or angle of the ladder be?

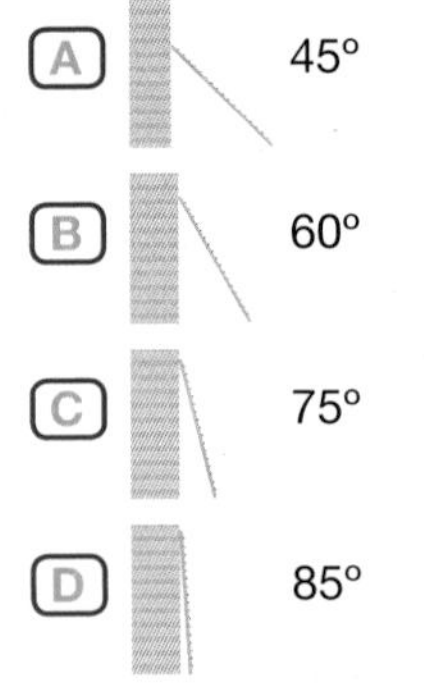

- (A) 45°
- (B) 60°
- (C) 75°
- (D) 85°

Answers: 14.17 = A 14.18 = C 14.19 = C 14.20 = C 14.21 = A 14.22 = C

14.23

To ensure your safety when carrying out an inspection at height, mobile tower scaffolds must only be erected by:

A someone who has the instruction book

B someone who is trained, competent and authorised

C advanced scaffolders

D a worker of the hire company

14.24

As part of a feasibility study you are planning an inspection of roof trusses inside a factory building. What is the recommended maximum height for a free-standing mobile tower when used indoors?

A There is no height restriction

B Three lifts

C As specified by the manufacturer

D Three times the longest base dimension

14.25

After gaining access to the platform of a mobile tower that has its wheels locked, the first thing you should do is:

A check that the tower's brakes are locked on

B check that the tower has been correctly assembled

C close the access hatch to stop people or equipment from falling

D check that the tower does not rock or wobble

14.26

An outdoor tower scaffold has stood overnight in high winds and heavy rain. What should you consider before the scaffold is used?

A That the brakes still work

B Tying the scaffold to the adjacent structure

C That the scaffold is inspected by a competent person

D That the platform hatch still works correctly

14.27

A person is going to be working on a mobile tower but will not be erecting, altering or dismantling it. What training should they have?

A They need the same level of training as a person erecting, altering or dismantling

B They do not need any training

C They should be briefed on the safe use and hazards of working on a mobile tower

D They should be in possession of the manufacturer's instructions

Answers: 14.23 = B 14.24 = C 14.25 = C 14.26 = C 14.27 = C

14.28

How will you know the maximum weight or number of people that can be lifted safely on a mobile elevating work platform (MEWP)?

A The weight limit is reached when the platform is full

B It will say on the Health and Safety Law poster

C You will be told during site induction

D From an information plate fixed to the machine

14.29

When is it safe to use a mobile elevating work platform (MEWP) on soft ground?

A When the ground is dry

B When the machine can stand on scaffold planks laid over the soft ground

C When stabilisers or outriggers can be deployed onto solid ground

D Never

14.30

If someone is working from a cherry picker, they should attach their safety lanyard to a:

A strong part of the structure that they are working on

B designed anchorage point inside the platform

C secure point on the boom of the machine

D scaffold guard-rail

14.31

A mobile elevating work platform (MEWP) is being used to carry out work at height. What is the only circumstance in which it is acceptable to lower the platform using the ground-level controls?

A When the person using the ground-level controls is competent to do so

B In an emergency

C If the person working on the platform needs to step off the MEWP to gain access to the high-level work area

D If the person working on the platform needs both hands free to carry out the job in hand

14.32

You have to carry out an inspection at height, using a cherry picker. You would NOT clip yourself to the machine using a restraint lanyard if the work involved:

A any type of roof work

B working over or near to deep water

C clambering from the machine on to the structure

D standing on the mid guard-rail to carry out the inspection

Answers: 14.28 = D 14.29 = C 14.30 = B 14.31 = B 14.32 = B

14.33

Your organisation's policy is to avoid walking on fragile roof materials. A common example of fragile roof material is:

- A asphalt felt roof
- B asbestos cement sheets or plastic rooflights
- C raised seam roofs
- D single-ply membrane

14.34

You are inspecting a flat roof. What is the best way to stop yourself and others from falling over the edge?

- A Have a large warning sign placed at the edge of the roof
- B Ask someone to keep watch and to shout out when someone gets too close to the edge
- C Ask for the edge to be protected with a guard-rail and toe-board
- D Ask for red and white tape to mark the edge

14.35

What is the best way to stop people falling through voids, holes or fragile roof panels?

- A Tell everyone where the dangerous areas are
- B Covers, secured in place, that can take the weight of a person and add warning signage
- C Cover them with netting
- D Mark the areas with red and white tape

14.36

What does this sign mean?

- A Load bearing roof. OK to stand on surface but not any rooflights
- B Fragile roof. Take care when walking on roof surface
- C Fragile roof. Do not stand directly on roof but use fall protection measures
- D Load-bearing roof. Surface can be slippery when wet

14.37

To ensure your safety and that of the site workers, what should the contractor include in a safety method statement for working at height? Give THREE answers.

- A The cost of the job and time it will take
- B The sequence of operations and the equipment to be used
- C How much insurance cover will be required
- D How falls are to be prevented
- E Who will supervise the job on site

Answers: 14.33 = B 14.34 = C 14.35 = B 14.36 = C 14.37 = B, D, E

14.38

When work is being carried out above public areas, your first consideration should be to:

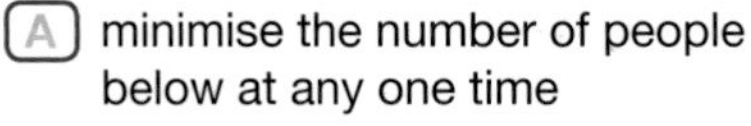

A minimise the number of people below at any one time

B provide alternative routes for the public and keep them away from the area

C let the public know what you are doing

D provide netting to prevent falling objects

14.39

Which of the following provides the public with adequate protection from falling objects?

A Clear warning signs

B A worker in a hi-vis vest standing on the footway to warn people

C Viewing panels in the site hoarding

D A fan or scaffold 'tunnel' over the footway

14.40

All work at height must be:

A risk assessed and properly planned

B only undertaken by scaffold contractors

C carried out as quickly as possible

D suspended if the forecast wind speed is above Force 2

14.41

How should access be controlled, if people are working in a riser shaft?

A By a site security operative

B By those who are working in it

C By the main contractor

D By a permit to work system

14.42

You are working at height taking site measurements, but the securing cord for a safety net is in your way. What should you do?

A Untie the cord, carry out your work and tie it up again

B Untie the cord, but ask the contractor to re-tie it when you have finished

C Tell the contractor that you are going to untie the cord

D Leave the cord alone and report the problem to the contractor

14.43

While carrying out a site visit you see someone who is working above a safety net system that has a damaged net. What should you do?

A Ask them to work somewhere away from the damaged area of net

B Ask them to stop work and report it to the site manager

C Ask them to tie the damaged edges together using the net test cords

D Ask them to go and see if they can get hold of a harness and lanyard

Answers: 14.38 = B 14.39 = D 14.40 = A 14.41 = D 14.42 = D 14.43 = B

14.44

A scaffold guard-rail must be removed to allow you to carry out a survey. You are not a scaffolder. Can you remove the guard-rail?

A Yes, if you put it back as soon as you have finished

B Yes, if you put it back before you leave site

C No, only a scaffolder can remove the guard-rail but you can put it back

D No, only a scaffolder can remove the guard-rail and put it back

14.45

To ensure your safety when using a scaffold to carry out an inspection at height, a competent person must routinely inspect the scaffold:

A before it is first used and then at intervals not exceeding seven days

B only after it has been erected

C after it has been erected and then at monthly intervals

D after it has been erected and then at intervals not exceeding 10 days

14.46

When can someone who is not a scaffolder remove parts of a scaffold?

A If the scaffold is not more than two lifts in height

B As long as a scaffolder refits the parts after the work has finished

C Never, only competent scaffolders can remove the parts

D Only if it is a tube and fittings scaffold

14.47

You need to use a ladder to get to a scaffold platform. Which of these statements is true?

A It must be tied and extend about five rungs above the platform

B All broken rungs must be clearly marked

C It must be wedged at the bottom to stop it slipping

D Two people must be on the ladder at all times to provide stability

14.48

If a scaffold is not complete, which of the following actions should be taken by the site manager?

A Make sure the scaffolders complete the scaffold

B Tell all operatives not to use the scaffold

C Use the scaffold with care and display a warning notice

D Prevent access to the scaffold by unauthorised people

Answers: 14.44 = D 14.45 = A 14.46 = C 14.47 = A 14.48 = D

15.1

You are in a deep trench and start to feel dizzy. What should you do?

A Ask others if they feel dizzy, if they don't then carry on for five minutes

B Have a drink – it's the first sign of dehydration

C Make sure that you and any others get out quickly and report it

D Sit down, put your head between your knees and take deep breaths to get some oxygen back into your system

15.2

Which of these is NOT a hazard in a confined space?

A Toxic gas

B A lack of carbon dioxide

C A lack of oxygen

D Flammable or explosive gas

15.3

Why is methane gas dangerous in confined spaces? Give TWO answers.

A It can explode

B It makes you hyperactive

C You will not be able to see because of the dense fumes

D It makes you dehydrated

E You will not have enough oxygen to breathe

15.4

You are in a confined space. If the level of oxygen drops:

A your hearing could be affected

B there is a high risk of fire or explosion

C you could become unconscious

D you might get dehydrated

15.5

You are working in a confined space when you notice the smell of bad eggs. This smell is a sign of:

A hydrogen sulphide

B oxygen

C methane

D carbon dioxide

15.6

You need to walk through sludge at the bottom of a confined space. Which of these is NOT a hazard?

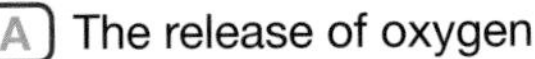

A The release of oxygen

B The release of toxic gases

C Slips and trips

D The release of flammable gases

Answers: 15.1 = C 15.2 = B 15.3 = A, E 15.4 = C 15.5 = A 15.6 = A

15.7

You are in a confined space when the gas alarm sounds. You have no respiratory protective equipment (RPE). What should you do?

- A Reset the gas alarm and see if it goes off again
- B Get out of the confined space quickly, while warning others
- C Reset the gas alarm and test another area in the confined space
- D Wait for one minute and see if the gas alarm cancels

15.8

As part of a site survey you are required to work in a confined space. How should the air be checked?

- A Unsafe atmospheres have a particular odour so someone should go in and smell the air
- B No-one should enter until the air has been tested with the appropriate gas detection meter
- C Warning signs will indicate the presence of an unsafe atmosphere
- D Unsafe atmospheres have little oxygen so the air should be tested with a flame to see if it stays alight

15.9

What danger is created by excessive oxygen in a confined space?

- A Increase in breathing rate of workers
- B Increased flammability of combustible materials
- C Decreased working time inside work area
- D False sense of security

15.10

Guard-rails should be placed around the top of an excavation to prevent:

- A plant from toppling into the excavation
- B anyone falling into the trench and being injured
- C the sides of the trench from collapsing
- D material from spoilt heaps falling into the excavation

15.11

You are standing near a deep trench. A lorry backs up to the trench and the engine is left running. What should you do?

- A Put on ear defenders to cut out the engine noise
- B Ignore the problem, the lorry will soon drive away
- C Look to see if there is a toxic gas meter in the trench
- D Get everyone out of the trench quickly

Answers: 15.7 = B 15.8 = B 15.9 = B 15.10 = B 15.11 = D

15.12

An excavation must be supported if:

A it is more than 5 m deep

B it is more than 1.2 m deep

C there is a risk of the sides falling in

D any buried services cross the excavation

15.13

You are looking at an excavation. If you see the side supports move, you should first:

A keep watching to see if they move again

B make sure that everyone working in the excavation gets out quickly

C do nothing as slight movement in the supports is quite normal

D move to another part of the excavation

15.14

Which of the following is a significant hazard when excavating alongside a building or structure?

A Undermining or weakening the foundations of the building

B Noise and vibration affecting the occupiers of the building

C Ground water could enter the excavation

D Damage to the surface finish of the building or structure

15.15

When is it advisable to take precautions to prevent the fall of persons, materials or equipment into an excavation?

A At all times

B When the excavation is 2 m deep or more

C When more than five people are working in the excavation

D When there is a risk from an underground cable or other service

15.16

You are inspecting an excavation into which dumpers are tipping material. What would you expect to see to prevent dumpers from falling into or damaging the edge of the excavation?

A Dumpers kept 5 m away from the excavation

B Stop blocks provided, parallel to the trench, appropriate to the vehicle's wheel size

C Dumper drivers required to judge the distance carefully or given stop signals by another person

D Cones or signage erected to indicate safe tipping point

Answers: 15.12 = C 15.13 = B 15.14 = A 15.15 = A 15.16 = B

15.17

The current Construction (Design and Management) Regulations require a supported excavation to be inspected by a competent person:

- A every seven days
- B at the start of the shift when the work is to be carried out
- C once a month
- D when it is more than 2 m deep

15.18

Which of these is the most accurate way to locate buried services?

- A Existing service drawings
- B Trial holes
- C Survey drawings
- D Architect drawings

15.19

Which piece of equipment may need to be used with a cable avoidance tool (CAT) in order to detect cables?

- A Insulated shovel or spade
- B Signal generator (genny)
- C Excavator bucket with no teeth
- D Gas detector

15.20

What must happen each time before a shift starts work in an excavation?

- A The workers should tighten any loose supports
- B A competent person must inspect the excavation
- C The workers should go down and pump out any rainwater
- D The workers should go down and check all is OK and report back to the supervisor

15.21

What is the safe way to get into a deep excavation?

- A Climb down a ladder
- B Use the buried services as steps
- C Climb down the shoring or trench supports
- D Climb down a secured ladder which extends 5 m past the stepping on point

15.22

If you need to work in a confined space, one duty of the top man is to:

- A tell you how to work safely in confined spaces
- B enter the confined space if there is a problem

- C start the rescue plan if needed
- D supervise the work in the confined space

Answers: 15.17 = B 15.18 = B 15.19 = B 15.20 = B 15.21 = D 15.22 = C

15.23

The best way to avoid the potential for someone becoming trapped in an excavation is to:

A eliminate the need for anyone to go into it

B check the contractor's method statement

C review the last excavation inspection record

D go down in a cage suspended from a crane

15.24

If there is the potential for work to be carried out in a confined space the FIRST consideration should be, can it be:

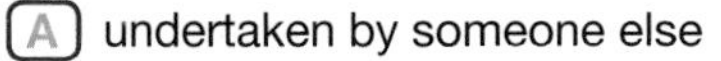
A undertaken by someone else

B avoided where possible

C managed by limiting the amount of time people are in it

D controlled with radios or mobile phones

15.25

Work in a confined space usually needs three safety documents – a risk assessment, a method statement and a:

A permit to work

B hot work permit

C COSHH assessment

D health and safety policy

15.26

You are required to undertake a survey in what you suspect might be a confined space. What should be your first consideration?

A Make sure that you have the correct equipment to test the atmosphere

B Make sure that you have a colleague with you

C Check if the information you need can be obtained remotely to avoid entering the space

D Follow a safe system of work for entering the confined space, including emergency arrangements

15.27

You are arranging for an inspection in a confined space to be carried out. Which are the TWO most important safety requirements?

A Completing a permit to work form

B Informing your office before entering

C Entering slowly and breathing shallowly until you are sure the air is good to breath

D Ensuring that a safe system of work has been identified and is followed

E Making sure that emergency rescue arrangements are in place

Answers: 15.23 = A 15.24 = B 15.25 = A 15.26 = C 15.27 = D, E

15.28

Before planning for anyone to enter a confined space, following the principles of prevention what should be the first consideration of the contractor's responsible person?

A Has the atmosphere in the confined space been tested?

B Has a safe means of access and egress been established?

C Is there an alternative method of doing the work?

D Have all who intend to enter the confined space been properly trained?

Answers: 15.28 = C

Environment

16.1

You become aware that operatives have destroyed the nest of nesting birds during a refurbishment job. Which of the following is a possible outcome?

A. A visit from an RSPCA inspector

B. A prohibition notice issued by the Health and Safety Executive (HSE)

C. A caution by the local police force

D. Prosecution by the Environment Agency

16.2

What is the legal duty of the site manager in relation to a non-native invasive species of plant, such as Japanese knotweed, which is discovered on site?

A. To ensure it is transplanted in a part of the site where it will not be disturbed in the future

B. To prevent the spread of the plant

C. To inform the Health and Safety Executive (HSE) of its presence

D. To leave it undisturbed as it is protected by law

16.3

Under Environmental Law which statement is true about protected species?

A. It is not illegal to disturb or destroy habitats of protected species if you didn't know they were there

B. It is only illegal to disturb or destroy habitats of protected species if you do it deliberately

C. You can disturb or destroy habitats of protected species if they get in the way of building work

D. It is illegal to disturb or destroy habitats of protected species

16.4

You discover a bird on a nest where you need to work. What should you do?

A. Cover it with a bucket

B. Move it, do your work and then put it back

C. Leave it alone and inform your supervisor

D. Scare it away

16.5

How should contractors dispose of hazardous waste from site?

A. Put it in any skip on site, as long as its less than 10% of the total volume

B. Separately and in accordance with the site waste management plan

C. It should be buried on site

D. It should be taken to the nearest local authority waste tip

Answers: 16.1 = D   16.2 = B   16.3 = D   16.4 = C   16.5 = B

16.6

A skip-lorry driver hands you a copy of the Hazardous Waste Consignment Note with Section C (carrier's details) signed. If you are responsible for the waste, what is the first thing you need to do with it?

(A) Refuse to take the note as you do not need a copy

(B) Take the note and file it for future reference

(C) Check that the details in sections A, B and C are correct and sign Section D (producer's details)

(D) As the waste is on his truck it is not your responsibility to sign the note

16.7

You are in charge of a waste compound and someone comes to you with a half empty paint tin. Is this hazardous waste?

(A) Yes, all paint is hazardous

(B) Yes, if it has 'hazardous symbols' on the packaging

(C) Yes, but only if the paint was still liquid

(D) No, paint is never hazardous

16.8

You see that a skip on your site has a few asbestos tiles mixed into the waste. Does this make the whole skip hazardous waste?

(A) No, because the asbestos is only a very small part of the waste

(B) No, because bonded asbestos is not dangerous to health

(C) Yes, but only if the asbestos is more than 10% of the skip

(D) Yes, because any quantity of asbestos is hazardous regardless of how big the skip is

16.9

If you see environmental incidents and near misses while on site when should you report these?

(A) Never, this is a problem for the contractor

(B) During your next break

(C) As soon as practical

(D) At the end of the day just before you leave site

Answers: 16.6 = C 16.7 = B 16.8 = D 16.9 = C

16.10

A full 200-litre drum carrying either of these symbols has toppled over and the whole content has seeped into the ground. Which of the following agencies should be informed?

A The Health and Safety Executive (HSE)

B The EU Authority on Chemical Safety

C The Environment Agency

D The Health Protection Agency

16.11

When installing a fuel-oil storage area, the contractor must include:

A water type fire extinguishers

B suitable bunding

C hand-washing facilities

D a porous ground surface to absorb spillage

16.12

When setting up a fuel storage tank on site, a spillage bund must have a minimum capacity of the contents of the tank, plus:

A 10%

B 15%

C 20%

D 25%

16.13

On a contaminated land remediation project, which of the following would you expect to be in place to avoid contamination of the surrounding area?

A Warning signs that state that visitors are excluded from the site

B Overalls for all visitors

C Adequate provision for vehicle wheel washing

D Respiratory protective equipment (RPE) for all visitors

16.14

What final element is missing from this simple four point pollution incident response plan?
Stop – Contain – Clean up – ?

A Review

B Notify

C Take action

D Re-start work

16.15

When assessing pollution risk, which of the following should NOT be done prior to starting works on a construction site?

A Locate and identify surface water drains with blue paint

B Seal up all drains and gullies on site

C Identify the risk of pollution for work activities entering existing drains

D Inspect existing gullies, silt traps and oil separators

Answers: 16.10 = C 16.11 = B 16.12 = A 16.13 = C 16.14 = B 16.15 = B

16.16

Which TWO of the following are possible pollution risks that could result from excavation activities?

A Collapse of the sides

B Contaminated soils

C Pumping out of silty water

D Electrocution from buried services

E Disturbing protected species

16.17

Which TWO of the following should be undertaken to help improve a pollution incident response?

A Train workforce in the use of spill kits

B Always refuel using drip trays

C Practise the incident response by undertaking mock exercises

D Train one 'responsible person' in the use of spill kits

E Only use biodegradable fuels

16.18

Which TWO of the following will help to minimise dust from stockpiles of soil?

A Avoid moving materials when nearby residents are home

B Damping down the materials with water

C Seed the stockpile

D Regularly move/mix materials between stockpiles

E Have a supply of face masks suitable for nuisance dust

16.19

To prevent pollution to watercourses what TWO things should you consider to control surface water runoff from material stockpiles?

A Digging 'cut-off' trenches around the stockpile

B Directing the water runoff away from the watercourse to the nearest surface water drain

C Channelling the water runoff directly into the foul water sewer

D Making sure stockpiles are more than 215 m away from watercourses

E Installing silt fence around the stockpile

16.20

What TWO precautions must be taken to reduce the risk of water pollution from a concrete or mortar batching plant? It should be:

A at least 5 m from a watercourse

B at least 10 m from a watercourse

C sited on a designated impermeable area

D not allowed on site if water pollution is a possibility

E only used under the direct supervision of an Environment Agency inspector

Answers: 16.16 = B, C 16.17 = A, C 16.18 = B,C 16.19 = A, E 16.20 = B, C

16.21

A concrete washout container that contains a significant quantity of concrete 'washwaters', that have settled out and look clear, has to be emptied. Which TWO of the following may be possible to avoid pollution?

- A Pump the clear water into the foul sewer with permission from the sewage undertaker
- B Pump and spread the water across any grassed areas on site
- C Reuse the water as part of future concrete mixing on site
- D Pump the water into the surface water drain
- E Pump the water down the nearest road gully

16.22

Which of the following are TWO environmental reasons for preventing concrete, screed or mortar 'wash out water' from entering watercourses or underground aquifers?

- A It can change the colour of the water
- B It can change the pH balance of the water
- C It will improve the flow of the watercourse
- D It will pollute the water and could harm water wildlife
- E It may be part of the conditions of contract and a client requirement

16.23

A plant refuelling point is to be set up on a new site. Which of the following is the preferred method of fuel transfer?

- A Decanting from jerry cans using a funnel
- B Gravity feed from a bulk storage tank
- C Siphon the fuel by mouth
- D Pumped system with a nozzle fitted with an auto cut-off device

16.24

What is the Code for Sustainable Homes?

- A A new safety code for house building
- B Rules on building homes with extra features that are required to make it fit for the future
- C An advert for developers to charge more for their houses
- D Guidance on how to give a house a makeover

Answers: 16.21 = A, C 16.22 = B, D 16.23 = D 16.24 = B

16.25

What is sustainable development?

A Development that sorts out our needs now

B Development that both deals with our current needs and our future needs

C Development only about our future needs

D Development that is concerned with giving aid to foreign countries

16.26

Under Environmental Law, which statement is true?

A Companies AND individuals can be prosecuted if they do not follow the law

B It is illegal to disturb protected species or their habitats

C It is illegal to transport waste without a licence

D All of these answers

16.27

Which THREE of the following are ways to indicate that site management is complying with the Site Waste Management Plans Regulations?

A Operatives are informed of waste management issues during site induction

B Detailed records are kept of how and when waste was removed, who removed it and where it was taken

C Subcontractors deal with their own waste and keep their own records

D Wastes are separated into different types of materials

E Hazardous and non-hazardous waste are put into the same skip for separation at the waste transfer station

16.28

What is Landfill Tax?

A A tax on construction generally

B A tax on buying land to build on

C A tax on disposing of waste

D A tax to help local authorities to earn income

Answers: 16.25 = B 16.26 = D 16.27 = A, B, D 16.28 = C

16.29

Why do we need to separate waste plaster and plasterboard from other types of waste?

A Because they are worth money

B Because they can react with other landfill wastes and produce a toxic gas

C Because we are running out of plasterboard and need to reuse them

D Because the client wanted us to do so

16.30

Overnight someone places an old fridge in your general waste skip. What should you do?

A Make sure the fridge is covered with other waste.

B Take the fridge out of the skip and dispose of it separately

C Tell the skip driver that there is a fridge in the skip

D Break the fridge up in the skip

16.31

You are aware that a job has resulted in some plasterboard off-cuts. Can these go in with the general waste?

A Yes, because plasterboard is not hazardous waste

B Yes, because the boards will only be a small proportion of the skip content

C No, because plasterboard is hazardous waste

D No, because plasterboard should not be mixed with other wastes

16.32

Someone has turned up at the site and offered to take your surplus soil away for free. Which of the following is NOT a legal requirement?

A They are a registered waste carrier

B They are able to complete a duty of care note

C They have got a clean driving licence

D They will take the soil to an authorised site

Answers: 16.29 = B 16.30 = B 16.31 = D 16.32 = C

F

Specialist activities

17.1

Under the current Construction (Design and Management) Regulations, who has responsibility for appointing a competent CDM co-ordinator?

A All duty holders

B The client

C Any contractor

D The principal contractor

17.2

If a designer appoints another designer or a contractor, the current Construction (Design and Management) Regulations requires them to:

A agree production and payment terms

B ensure that they attend the site induction

C be sure that they carry out their duties under the regulations

D be satisfied that they are competent

17.3

Under the current Construction (Design and Management) Regulations, who has the legal responsibility for ensuring that project arrangements are in place for the allocation of sufficient resources?

A The CDM co-ordinator

B The client

C The lead designer

D The principal contractor

17.4

In what circumstances must a construction phase health and safety plan be prepared?

A They are required on every construction project

B Only if the client requires one

C Only if it is a contractual requirement

D Whenever the project is notifiable under the CDM Regulations

Answers: 17.1 = B 17.2 = D 17.3 = B 17.4 = D

17.5

In meeting the risk management objectives of the current Construction (Design and Management) Regulations, which statement summarises the best project attitude for improving health and safety on construction sites?

A Only the CDM co-ordinator can eliminate hazards

B Only the client has the budget to eliminate hazards

C Designers and contractors can eliminate significant hazards

D Designers identify hazards and contractors eliminate them

17.6

Initially, who is most likely to hold most information about the health and safety constraints of the site?

A The CDM co-ordinator

B The principal contractor

C The Health and Safety Executive (HSE)

D The client

17.7

Who is responsible for managing health and safety on construction sites?

A The CDM co-ordinator

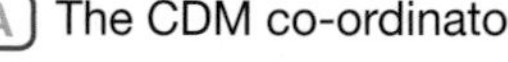

B The principal contractor

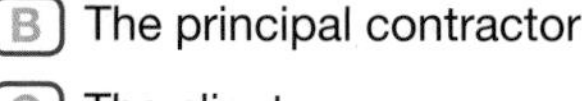

C The client

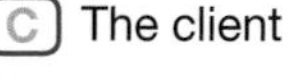

D The designer

17.8

The design team has agreed with the client to omit reference to asbestos in a work package, even though asbestos is present in the structure. In this circumstance the design team and the client have acted:

A reasonably practicably

B illegally

C legally

D with adequate regard

17.9

When visiting a structure built in the last 10 years where would you expect to find information about the risks associated with the completed structure?

A Construction phase plan

B Operation and maintenance manuals

C Original project specification

D Health and safety file

Answers: 17.5 = C 17.6 = D 17.7 = B 17.8 = B 17.9 = D

17.10

Under the current Construction (Design and Management) Regulations the duty to apply the principles of prevention rests with which duty holders?

A Designers and clients

B Contractors and principal contractors

C The CDM co-ordinator

D All duty holders

17.11

The current Construction (Design and Management) Regulations require contractors and designers to only commence work if:

A clients are supplied with information about the client's duties under the CDM Regulations

B their contracts with clients address the client's obligations under the CDM Regulations

C the Health and Safety Executive (HSE) is notified of the project before a design is prepared

D clients are aware of the client's duties under the CDM Regulations before design work is started

17.12

You plan to visit a site that you know is notifiable and you realise that a CDM co-ordinator has not been appointed. What should you do?

A Contact the client immediately and ensure that they are aware of their duties

B Work to the client's brief

C Complete all work within your brief and ignore the lack of a CDM co-ordinator

D Stop all work on the project immediately and notify the Health and Safety Executive (HSE)

17.13

What is the most effective way for designers to identify and communicate residual risks to the site management?

A Adding notes to the specification

B Supplying material safety data sheets

C Adding notes to drawings

D Notes in a bill of quantities

Answers: 17.10 = D 17.11 = D 17.12 = A 17.13 = C

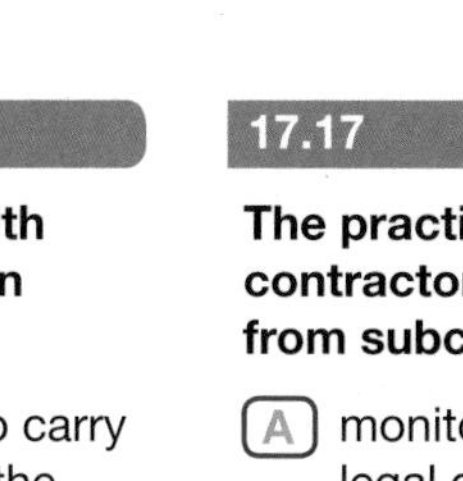

17.14

What is the purpose of the health and safety file on a construction project?

- (A) To assist people who have to carry out work on the structure in the future
- (B) To assist in the preparation of final accounts for the structure
- (C) To record the health and safety standards of the structure
- (D) To record the accident details

17.15

Which one of the following directly controls the way works are undertaken on site?

- (A) The principal contractor's health and safety policy statement
- (B) The local Health and Safety Executive (HSE) inspector
- (C) The principal contractor's construction phase health and safety plan
- (D) The Health and Safety at Work etc. Act 1974

17.16

In order to manage the safety of site visitors, which of the following documents must the principal contractor keep under review?

- (A) The health and safety file
- (B) Designer's risk assessments
- (C) Project programme
- (D) The construction phase health and safety plan

17.17

The practical way for the principal contractor to ensure co-operation from subcontractors is to:

- (A) monitor their works, explain their legal duties and help them comply
- (B) haul them over the coals when they do something wrong
- (C) withhold payments when they do not perform satisfactorily
- (D) wait to make comments at the next progress meeting

17.18

Under the current Construction (Design and Management) Regulations, which of the following must the principal contractor ensure is specifically provided before allowing any demolition work to commence?

- (A) A construction phase safety plan
- (B) The arrangements for demolition recorded in writing
- (C) A generic risk assessment
- (D) A pre-tender health and safety plan

17.19

To whom should the CDM co-ordinator pass the health and safety file on completion of the construction project?

- (A) The Association for Project Safety
- (B) The client
- (C) The Health and Safety Executive (HSE)
- (D) The designer

Answers: 17.14 = A 17.15 = C 17.16 = D 17.17 = A 17.18 = B 17.19 = B

17.20

Where must you be able to find the name and address of the client, CDM co-ordinator and principal contractor?

A The construction phase health and safety plan

B The Health and Safety Executive (HSE) Form F10 or the equivalent for the project

C Displayed on a project sign board near the welfare facilities

D The accident book

17.21

Where a project is notifiable under the current Construction (Design and Management) Regulations, what must be in place before construction work begins?

A Construction project health and safety file

B Construction phase health and safety plan

C Construction project plan

D Construction contract agreement

17.22

Where a project is notifiable under the current Construction (Design and Management) Regulations, who is responsible for ensuring notification to the Health and Safety Executive (HSE) of the project?

A Client

B Designer

C CDM co-ordinator

D Principal contractor

17.23

If a new build project is notifiable under the current Construction (Design and Management) Regulations, the client must ensure that construction does not start until:

A the construction phase health and safety plan is in place

B a site manager has been employed to take charge

C the Health and Safety Executive (HSE) has given permission

D the health and safety file is in place

17.24

During the construction phase, the CDM co-ordinator has responsibilities for which TWO activities under the current Construction (Design and Management) Regulations?

A Ensuring co-operation between designers and the principal contractor

B Appointing a competent and adequately resourced designer

C Deciding which construction processes are to be used

D Ensuring that relevant pre-construction information is identified and collected

E The on-going monitoring of site safety throughout the construction phase

Answers: 17.20 = B 17.21 = B 17.22 = C 17.23 = A 17.24 = A,D

17.25

Where a project is notifiable under the current Construction (Design and Management) Regulations, who is responsible for preparing the construction phase health and safety plan?

A The principal contractor

B The client

C A contractor tendering for the project

D The CDM co-ordinator

17.26

Except for work being carried out for a domestic client, under the current Construction (Design and Management) Regulations, in which of the following situations must the Health and Safety Executive (HSE) be notified of a project?

A Where the work will last more than 30 days or more than 500 person-days

B Where the building and construction work will last more than 300 person-days

C Where the work will last more than 50 days or more than 300 person-days

D Where the work will be done by more than 30 people or last more than 500 hours

17.27

Under the current Construction (Design and Management) Regulations, who is responsible for initially making pre-construction information available?

A The CDM co-ordinator

B The principal contractor

C The client

D The client's agent

17.28

Before starting any construction work lasting more than 30 days, or 500 person-days, which of the following must be done?

A The local authority must be informed on a Form F9

B The health and safety file must be handed to the client

C The client must prepare a pre-tender health and safety plan

D The Health and Safety Executive (HSE) must be notified

17.29

Where a project is notifiable under the requirements of the current Construction (Design and Management) Regulations, what has to be displayed on a construction site?

A Notice of application to erect hoardings

B Notice of the principal contractor's health and safety policy

C Form F10 (Rev) or a notice carrying specified information

D A statement by the client

Answers: 17.25 = A 17.26 = A 17.27 = C 17.28 = D 17.29 = C

18.1

If there are any doubts as to a building's stability, a demolition contractor should consult:

A another demolition contractor

B a structural engineer

C a Health and Safety Executive (HSE) factory inspector

D the company safety adviser

18.2

Which one of the following is an effective way of ensuring good standards of health and safety on a demolition project?

A Checking the contractor's method statement

B Selecting a competent demolition contractor

C Ensuring operatives use personal protective equipment (PPE) as necessary

D All of these answers

18.3

Every demolition contractor undertaking demolition operations must first appoint:

A a competent person to supervise the work

B a subcontractor to strip out the buildings

C a safety officer to check on health and safety compliance

D a quantity surveyor to price the extras

18.4

What action should be taken if the contractor discovers unlabelled drums or containers on site?

A Put them in the nearest waste skip

B Ignore them. They will get flattened during the demolition

C Stop work until they have been safely dealt with

D Open them and smell the contents

18.5

You are required to visit a demolition site. What is the most common source of high levels of lead in the atmosphere during demolition work?

A Stripping lead sheeting

B Cutting lead-covered cable

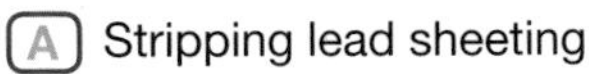

C Cold cutting fuel tanks

D Hot cutting coated steel

18.6

After exposure to lead, what precautions should you take before eating or drinking?

A Wash your hands and face

B Do not smoke

C Change out of dirty clothes

D Rinse your mouth with clean water

Answers: 18.1 = B 18.2 = D 18.3 = A 18.4 = C 18.5 = D 18.6 = A

18.7

When asbestos material is suspected in buildings to be demolished, what is the FIRST priority?

A A competent person carries out an asbestos survey

B Notify the Health and Safety Executive (HSE) of the possible presence of asbestos

C Remove and dispose of the asbestos

D Employ a licensed asbestos remover

18.8

What do the letters SWL stand for?

A Satisfactory working limit

B Safe working level

C Satisfactory weight limit

D Safe working load

18.9

Which of the following is true as regards the safe working load of a piece of equipment?

A It must never be exceeded

B It is a guide figure that may be exceeded slightly

C It may be exceeded by 10% only

D It gives half the maximum weight to be lifted.

18.10

What should be clearly marked on all lifting gear?

A Date of manufacture

B Name of maker

C Date next test is due

D Safe working load

18.11

What action should be taken by the contractor if a wire rope sling is defective?

A Do not use it and make sure that no-one else can

B Only use it for up to half its safe working load

C Put it to one side to wait for repair

D Only use it for small lifts under 1 tonne

18.12

With regard to mobile plant, what safety feature is provided by FOPS?

A The speed is limited when tracking over hard surfaces

B The machine stops automatically if the operator lets go of the controls

C The operator is protected from falling objects

D The reach is limited when working near to live overhead cables

Answers: 18.7 = A 18.8 = D 18.9 = A 18.10 = D 18.11 = A 18.12 = C

18.13

What should a contractor do if they discover underground services not previously identified?

A. Fill in the hole and say nothing to anyone

B. Stop work until the situation has been resolved

C. Cut the pipe or cable to see if it's live

D. Get the machine driver to dig it out

18.14

Which is the safest method of demolishing brick or internal walls by hand?

A. Undercut the wall at ground level

B. Work across in even courses from the ceiling down

C. Work from the doorway at full height

D. Cut down at corners and collapse in sections

18.15

Who should be consulted before demolition is carried out near to overhead cables?

A. The Health and Safety Executive (HSE)

B. The fire service

C. The electricity supply company

D. The land owner

18.16

Where would you find the intended method of controlling identified hazards on a demolition project?

A. The demolition tool box talk

B. The demolition plan

C. The pre-tender health and safety plan

D. The construction phase health and safety plan

18.17

Before a contractor enters large, open-topped tanks, what is the most important thing they should obtain?

A. A ladder for easy access

B. A valid permit to work

C. An operative to keep watch

D. A gas meter to detect any gas

18.18

Before carrying out the demolition cutting of fuel tanks what should be obtained?

A. A gas free certificate

B. An isolation certificate

C. A risk assessment

D. A COSHH assessment

Answers: 18.13 = B 18.14 = B 18.15 = C 18.16 = B 18.17 = B 18.18 = A

18.19

Which TWO of the following documents refer to the specific hazards associated with demolition work in confined spaces?

- A Safety policy
- B Permit to work
- C Risk assessment
- D Scaffolding permit
- E Hot work permit

Answers: 18.19 = B, C

19.1

The legionella bacteria that cause Legionnaire's disease are most likely to be found in which of the following?

A A boiler operating at a temperature of 80°C

B A shower hose outlet

C A cold water storage cistern containing water at 10°C

D A WC toilet pan

19.2

How are legionella bacteria passed on to humans?

A Through fine water droplets, such as sprays or mists

B By drinking dirty water

C Through contact with the skin

D From other people when they sneeze

19.3

Which of the following is most likely to result in those who work with sheet lead having raised levels of lead in their blood?

A By them not using the correct respirator

B By not washing their hands before eating

C By not changing out of their work clothes

D By them not wearing safety goggles

19.4

Apart from the cylinders used in gas-powered fork-lift trucks, you should never see liquefied petroleum gas cylinders placed on their sides during use because:

A it would give a faulty reading on the contents gauge, resulting in flashback

B air could be drawn into the cylinder, creating a dangerous mixture of gases

C the liquid gas would be at too low a level to allow the torch to burn correctly

D the liquid gas could be drawn from the cylinder, creating a safety hazard

19.5

What is the preferred method of checking for leaks when assembling liquefied petroleum gas equipment before use?

A Test with a lighted match

B Sniff the connections to detect the smell of gas

C Listen to hear for escaping gas

D Apply leak detection fluid to the connections

19.6

What is the colour of propane gas cylinders?

A Black

B Maroon

C Red/orange

D Blue

Answers: 19.1 = B 19.2 = A 19.3 = B 19.4 = D 19.5 = D 19.6 = C

19.7

Which of the following makes it essential for contractors to take great care when handling oxygen cylinders?

A They contain highly flammable compressed gas

B They contain highly flammable liquid gas

C They are filled to extremely high pressures

D They contain poisonous gas

19.8

What is the colour of an acetylene cylinder?

A Orange

B Black

C Green

D Maroon

19.9

Which of the following is the safest place to store oxyacetylene gas welding bottles when they are not in use?

A Outside in a special secure storage compound

B In company vehicles

C Inside the building in a locked cupboard

D In the immediate work area, ready for use the next day

19.10

When observing oxyacetylene welding equipment being used on site, the bottles should be:

A laid on their side

B stood upright

C stood upside down

D angled at 45°

19.11

What item of personal protective equipment (PPE), from the following list, should be used when oxyacetylene-welding?

A Ear defenders

B Clear goggles

C Green-tinted goggles

D Dust mask

19.12

When working in an area where fibreglass roof insulation is being handled, in addition to safety boots and helmet, which of the following items of personal protective equipment (PPE) should be worn?

A Gloves, face mask and eye protection

B Rubber apron, eye protection and ear defenders

C Ear defenders, face mask and knee pads

D Barrier cream, eye protection and face mask

Answers: 19.7 = C 19.8 = D 19.9 = A 19.10 = B 19.11 = C 19.12 = A

19.13

The reason for carrying out temporary continuity bonding before removing and replacing sections of metallic pipework is to:

A provide a continuous earth for the pipework installation

B prevent any chance of blowing a fuse

C maintain the live supply to the electrical circuit

D prevent any chance of corrosion to the pipework

19.14

You arrive to carry out a site inspection that involves using ladder access to a roof. You notice the ladder has been painted. You should:

A only use the ladder if it is made of metal

B only use the ladder if it is made of wood

C only use the ladder if wearing rubber-soled boots to prevent slipping

D not use the ladder, and report the matter to the site manager

Answers: 19.13 = A 19.14 = D

20.1

What should the site manager do for the safety of private motorists if transport leaving site is likely to deposit mud on the public road?

- (A) Have someone in the road to slow down the traffic
- (B) Employ an on-site method of washing the wheels of site transport
- (C) Employ a mechanical road sweeper
- (D) Have someone hosing down the mud in the road

20.2

From a safety point of view, diesel must not be used to prevent asphalt sticking to the bed of lorries because:

- (A) it will create a slipping hazard
- (B) it will corrode the bed of the lorry
- (C) it will create a fire hazard
- (D) it will react with the asphalt, creating explosive fumes

20.3

When kerbing works are being carried out, which method should be used for getting kerbs off the vehicle?

- (A) Lift them off manually using the correct technique
- (B) Push them off the back
- (C) Use mechanical means, such as a JCB fitted with a grab
- (D) Ask your workmate to give you a hand

20.4

What are TWO effects of under-inflated tyres on the operation of a machine?

- (A) It decreases the operating speed of the engine
- (B) It leads to instability of the machine
- (C) It causes increased tyre wear
- (D) It decreases tyre wear
- (E) It increases the operating speed of the engine

20.5

Which of the following is true as regards the safe working load of lifting equipment, such as a cherry picker, lorry loader or excavator?

- (A) It must never be exceeded
- (B) It is a guide figure that may be exceeded slightly
- (C) It may be exceeded by 10% only
- (D) It gives half the maximum weight to be lifted

20.6

Which of the following checks should the operator of a mobile elevating work platform (MEWP), for example a cherry picker, carry out before using it?

- (A) Check that a seatbelt is provided for the operator
- (B) Check that a roll-over cage is fitted
- (C) Drain the hydraulic system
- (D) Check that emergency systems operate correctly

Answers: 20.1 = B 20.2 = A 20.3 = C 20.4 = B, C 20.5 = A 20.6 = D

20.7

In which of the following circumstances would it NOT be safe to use a cherry picker for working at height?

A When a roll-over cage is not fitted

B When the ground is uneven and sloping

C When weather protection is not fitted

D When the operator is clipped to an anchorage point in the basket

20.8

Mobile works are being carried out by day. A single vehicle is being used. What must be conspicuously displayed on or at the rear of the vehicle?

A Road narrows (left or right)

B A specific task warning sign (for example, gully cleaning)

C A 'keep left/right' arrow

D A 'roadworks ahead' sign

20.9

What action is required when a vehicle fitted with a direction arrow is travelling from site to site?

A Point the direction arrow up

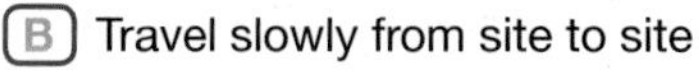

B Travel slowly from site to site

C Point the direction arrow down

D Cover or remove the direction arrow

20.10

When should the amber flashing beacon fitted to a vehicle be switched on?

A At all times

B When travelling to and from the depot

C When the vehicle is being used as a works vehicle

D Only in poor visibility

20.11

Lifting equipment for carrying persons, for example a cherry picker, must be thoroughly examined by a competent person every:

A 6 months

B 12 months

C 18 months

D 24 months

20.12

When undertaking a site survey on a dual carriageway with a 60 mph-speed-limit what is the minimum standard of hi-vis clothing that must be worn?

A Reflective waistcoat

B Reflective long-sleeved jacket

C Reflective sash

D None

Answers: 20.7 = B 20.8 = C 20.9 = D 20.10 = C 20.11 = A 20.12 = B

20.13

Why is it necessary to wear hi-vis clothing when working on roads?

A So road users and plant operators can see you

B So that your colleagues can see you

C Because you were told to

D Because it will keep you warm

20.14

When providing portable traffic signals on minor rural roads used by cyclists and horse riders, what action should be taken by the contractor?

A Locate the signals at bends in the road

B Allow more time for slow-moving traffic by increasing the 'all red' phase of the signals

C Operate the signals manually

D Use 'stop/go' boards only

20.15

Why should temporary signing be removed when works are complete?

A To get traffic flowing

B It is a legal requirement

C To allow the road to be opened fully

D To reuse signs on new jobs

20.16

When should installed signs and guarding equipment be inspected?

A After it has been used

B Once a week

C Before being used

D Regularly and at least once every day

20.17

Signs placed on footways must be located so that they:

A block the footway

B can be read by site personnel

C do not create a hazard for pedestrians

D can be easily removed

20.18

When drivers who are approaching roadworks cannot easily see any advance roadwork signs because of poor visibility or other obstructions, the contractor should:

A place additional signs in advance of the works

B extend the safety zones

C extend the sideways clearance

D lengthen the lead-in taper

Answers: 20.13 = A 20.14 = B 20.15 = B 20.16 = D 20.17 = C 20.18 = A

20.19

What action is required where passing traffic may block the view of signs?

A Signs must be larger

B Signs must be duplicated on both sides of the road

C Signs must be placed higher

D Additional signs must be placed in advance of the works

20.20

In which TWO places would you find information on the distances for setting out the signs in advance of the works under different road conditions?

A In the 'Traffic Signs Manual' (Chapter 8)

B In the 'Pink Book'

C On the back of the sign

D In the specification for highway works

E In the Code of Practice ('Red Book')

20.21

Signs, lights and guarding equipment must be properly secured:

A with sacks containing fine, granular material set at a low level

B by roping them to concrete blocks or kerb stones

C to prevent them being stolen

D by iron weights suspended from the frame by chains or other strong material

20.22

If you are working after dark, is mobile plant exempt from the requirement to show lights?

A Yes, on all occasions

B Yes, if authorised by the site manager

C Only if they are not fitted to the machine as standard

D Not in any circumstances

Answers: 20.19 = B   20.20 = A, E   20.21 = A   20.22 = D

Further information

Preparing for the case studies

Construction is an exciting industry. There is constant change as work progresses to completion. As a result, the building site is one of the most dangerous environments to work in.

Everyone on site working together can avoid many of the accidents that happen. The free film *Setting out* shows what you and the site must do to stay healthy and safe at work.

To watch the film you can:

 go online at *www.cskills.org/settingout.*

This film is essential viewing for everyone involved in construction, and should be viewed before sitting the CITB-ConstructionSkills' *Health, safety and environment test.*

The content of the film is summarised here. These principles form the basis for the behavioural case studies included in the test from Spring 2012.

It is advisable to watch the film. However, for those unable to view it or who want to refresh themselves on the content, the full transcript is provided for your information.

Part 1: What you should expect from the construction industry

Your site and your employer should be doing all they can to keep you and your colleagues safe.

Before any work begins, the site management team will have been planning and preparing the site for your arrival. It's their job to ensure that you can do your job safely and efficiently.

Five things the site you are working on must do:

- ☑ know when you are on site
- ☑ give you a site induction
- ☑ give you site-specific information
- ☑ encourage communication
- ☑ keep you up to date and informed.

Part 2: What the industry expects of you

Once the work begins, it's up to every individual to take responsibility for carrying out the plan safely.

This means you should follow the rules and guidelines as well as being alert to the continuing changes on site.

Five things you must do:

- ☑ respect and follow the site rules
- ☑ safely prepare each task
- ☑ carry out each task responsibly
- ☑ know when to stop (if you think anything is unsafe)
- ☑ keep learning.

Setting out

What to expect from the industry and what the industry expects from you

Construction is an exciting industry. There is constant change as work progresses to completion.

As a result, the construction site is one of the most dangerous environments to work in.

Many accidents that occur on sites can be avoided. In this film you will find out what you and the site must do to stay healthy and safe at work.

Part one: What to expect from the industry

When you arrive for your first day on a site, it will not be the first day for everyone.

Before any work begins, the site management team will have been planning and preparing the site for your arrival.

It is *their* job to ensure that you can do *your* job safely and efficiently.

So what are five key things the site should do for you?

1. Your site must know when you are on site

When you arrive you should be greeted and welcomed by someone on site. If you are not then make your presence known to site management.

You need to know who is in charge and they need to know who is working on, or visiting, their site.

You may be asked to sign in or report to someone in charge when you arrive.
You should also sign out or let someone know if you are leaving.

2. Your site must give you an induction

Once you have introduced yourself, you will be given a site induction. This is a legal requirement to give you basic information so that you can work safely on site.

You may be asked to watch a video or look at a presentation. It is important that you understand what is said during the induction. If there is something you are not sure about, don't be afraid to ask for more information. If you have not been to the site for a while, you need to be sure that you are up to date. Check with site management whether you need a briefing or a further induction.

3. Your site must give you site-specific information

Whether you are just starting out or have decades of experience, every job is different. So it is very important that the site induction is specific to your site.

You should be told about any specific areas of danger and what site rules are in place to control these.

You will be told who the managers are on site and what arrangements are in place for emergencies.

You will also find out what to do it there is a fire or if you need to sound an alarm.

It may sound basic, but there must be good welfare facilities.

You must also be able to take a break somewhere that is warm and dry.

4. Your site must encourage communication

Evidence shows that there are fewer incidents and accidents on sites where workers are actively involved in health and safety. Your opinions and ideas are important so make them heard.

Whatever the size of your site there should be many opportunities to do this. For example, directly with managers through a daily briefing or through suggestion boxes. Managers should let you know how best to do this on your site.

5. Your site must keep you up to date

Construction sites are constantly changing and unplanned activities can be a major cause of accidents. The more up to date you are about what is happening on site, the more you can understand the dangers. The best sites keep their people informed on daily activities.

Is there a hazards board on your site – is it regularly updated?

Your site management should be telling you what's going on, on a regular basis.

Your site should be doing all it can to keep you and your colleagues safe. If it is not – say something and work together to make it better.

Part two: What to expect from the industry

Site management will have planned ahead to make work on-site as safe as possible for you. Once the work begins, it is up to every individual to take responsibility for carrying out the plan safely. This means following the guidelines set out and being alert to the continuing changes on site.

So what will the site expect from you?

1. You must respect site rules

Site rules are there to minimise the risk of particularly hazardous activities, such as moving vehicles and handling flammable substances.

Moving traffic on site is a major cause of accidents. Often rules will cover issues like walking through the site, where to park, and how to behave when you see a moving vehicle.

They may also tell you where you can smoke and remind you to tidy equipment away when it is not in use.

These might feel restrictive but they have been put in place for a reason. If that reason is not clear to you, ask for more information. Otherwise, follow the rules to stay safe.

2. You must safely prepare each task

Every task carried out on site is unique and will have its own dangers, for instance working at height or manual handling. The site management team will put in place a plan to avoid or minimise the dangers before the start of work. The plan may be written down in a risk assessment, a method statement or a task sheet.

This will tell you what to do, the skills needed, what to wear, what tools to use and what the dangers are.

You should contribute to the planning process using your experience and knowledge.

For example, before working at height, which is a high-risk activity, you must consider:

- ☑ Is the access suitable for what you need to do?
- ☑ Do you have the right tools?
- ☑ Do you have the right protective equipment, does it fit, and is it comfortable?

3. You must do each task responsibly

Once you are at work you must apply your training, skills and common sense to your tasks at all times.

For example, if you are building a wall, you may have to move heavy loads. But you must not put your health and body in danger. Make sure you move the load as safely as possible, as you should have been trained to do. And if you are not sure how to, then you must ask for advice.

Equally, it is important to be aware of the dangers to those around you. For example, if someone working with you is not wearing the correct PPE for a certain activity, tell them so.

Acting responsibly will benefit both you and your colleagues.

4. You must know when to stop

In our industry saying NO is not easy. We are fixers and doers. We CAN do.

However, a significant number of accidents on sites happen when people are doing things that they're not comfortable with.

For example, many workers have been harmed by not knowing how to identify asbestos. If you think there is any likelihood that there is asbestos present where you are working, you must stop work and seek advice.

If you are not properly trained, equipped, or briefed – or if the situation around you changes – the result could be an accident.

Trust your instincts. If things feel beyond your control or dangerous or, if you see someone else working unsafely, stop the work immediately and inform site management why you have done so.

You might prevent an injury or save a life. Your employer should be supportive if you do this because you have the right to say no and the responsibility to not walk by.

5. You must keep learning

If your job requires you to have specific training to enable you to do it safely then it is your employer's responsibility to provide it.

To really get the best out of your career, you should keep learning about developments in machinery, equipment, regulations and training.

This will not only give you greater confidence and understanding, it will ensure you remain healthy and safe.

In summary

Construction is so much more than bricks and mortar. The work we do improves the world around us. It's time for us to work together to build a safer and better industry.

Acknowledgements

CITB-ConstructionSkills wishes to acknowledge the assistance offered by the following organisations in the preparation of the question banks:

- ConstructionSkills (NI)
- Construction Employers Federation Limited (CEF NI)
- Construction Industry Confederation (CIC)
- Construction Plant-hire Association (CPA)
- Environment Agency
- Federation of Master Builders (FMB)
- Health and Safety Executive (HSE)
- Heating and Ventilating Contractors' Association (HVCA)
- Highways Agency
- Joint Industry Board for Plumbing Mechanical Engineering Services (JIB-PMES)
- Lift and Escalator Industry Association (LEIA)
- National Access and Scaffolding Confederation (NASC)
- National Demolition Training Group (NDTG)
- Scottish and Northern Ireland Joint Industry Board for the Plumbing Industry (SNIJIB)
- Strategic Forum for Construction Competence Working Group
- TunnelSkills
- UK Contractors Group (UKCG)
- Union of Construction, Allied Trades and Technicians (UCATT)
- Unison
- Unite

Notes

Notes

Notes